REWARD

Pre-intermediate

Student's Book

Simon Greenall

MACMILLAN
HEINEMANN
English Language Teaching

Map of the book

Lesson	Grammar and functions	Vocabulary	Skills and sounds
Progress check lessons 6 – 10	Revision	Word chains Compound nouns Categorising vocabulary	**Sounds:** syllable stress in words; /ð/ and /θ/; /ɒ/ and /əʊ/; friendly intonation **Writing:** predicting a story from questions **Speaking:** talking about past events; families
11 *How ambitious are you?* Talking about ambitions	Verb patterns (2): *to* + infinitive; *going to* for intentions, *would like to* for ambitions	Ambitions Verbs and nouns which go together	**Reading:** reading and answering a questionnaire **Writing:** writing a paragraph describing your ambitions using *because* and *so*
12 *English in the future* The role of the English language in the future of your country	*Will* for predictions	Jobs School subjects	**Listening:** listening for main ideas **Sounds:** syllable stress in words; /e/ and /eɪ/ **Speaking:** talking about the future of English **Writing:** writing a paragraph about what people think about the future of English
13 *Foreign travels* Planning a trip to South America	*Going to* for plans and *will* for decisions Expressions of future time	Equipment for travellers	**Listening:** listening for specific information **Speaking:** planning a trip
14 *In Dublin's fair city* Finding your way around town	Prepositions of place Asking for and giving directions	Town features Adjectives to describe bars	**Reading:** reacting to a passage **Listening:** listening for specific information **Sounds:** /ɑː/, /æ/ and /ʌ/ **Speaking:** giving directions around town
15 *An apple a day* Typical meals in different countries	Expressions of quantity (1): countable and uncountable nouns, *some* and *any*, *much* and *many*	Food and drink Meals	**Listening:** listening for specific information **Speaking:** talking about typical meals and food in different countries
Progress check lessons 11 – 15	Revision	Word maps Nouns from verbs and nouns from other nouns Noun suffixes for jobs	**Sounds:** weak syllables /ə/ ;/tʃ/ and /ʃ/; contrastive stress; polite intonation in questions **Speaking:** planning a lunch party for friends
16 *What's on?* Typical entertainment in different countries	Prepositions of time and place Making invitations and suggestions	Types and places of entertainment and related words	**Listening:** listening for specific information **Speaking:** talking about typical entertainment **Writing:** writing and replying to invitations
17 *Famous faces* Describing people	Describing appearance and character: *look like, be like*	Words to describe height, age, looks, build and character	**Listening:** listening for main ideas **Speaking:** describing people **Writing:** writing a letter describing your appearance
18 *Average age* Personal qualities at different ages	Making comparisons (1): comparative and superlative adjectives	Adjectives of character	**Reading:** reacting to a passage and comparing information in a passage with your own experience **Speaking:** talking about exceptional people **Writing:** writing sentences describing exceptional people
19 *Dressing up* Typical clothes in different countries	Making comparisons (2): *more than, less than, as...as*	Clothes Colours Personal categories for organising new vocabulary	**Reading:** reading for specific information **Sounds:** weak syllables /ə/ and /ɪ/; weak forms /ðən/, /əz/ and /frəm/; stress for disagreement **Listening:** listening for main ideas **Speaking:** talking about clothing
20 *Memorable journeys* A car journey across the USA	Talking about journey time, distance, speed and prices	Numbers Words to describe a long-distance journey by car	**Listening:** listening for specific information **Sounds:** syllable stress in numbers **Speaking:** talking about a memorable journey
Progress check lessons 16 – 20	Revision	International words Adjective suffixes Male and female words	**Sounds:** /ʊ/ and /uː/; /dʒ/; polite and friendly intonation **Speaking:** mutual dictation **Writing:** mutual dictation to recreate a story

Lesson	Grammar and functions	Vocabulary	Skills and sounds
21 *How are you keeping?* Your body and your health	Present perfect simple (1) for experiences	Parts of the body	**Reading:** reading and answering a questionnaire **Speaking:** talking about experiences
22 *What's new with you?* Talking about changes in your life	Present perfect simple (2) for past actions with present results	Political and social conditions	**Listening:** predicting; listening for specific information **Sounds:** linking of /v/ and /s/ endings before certain verbs **Writing:** writing a letter describing recent changes in your life
23 *It's a holiday* Important national or local events and festivals	Present perfect simple (3): *for* and *since*	Words to describe important events and festivals	**Listening:** listening for specific information **Sounds:** weak form /fə/ **Writing:** writing a paragraph describing an important national occasion
24 *Divided by a common language?* A comparison of British and American English	Defining relative clauses: *who, which/that* and *where*	American English words with different meanings in British English	**Speaking:** talking about useful types of English **Reading:** inferring **Sounds:** comparing American and British standard pronunciation; difference in specific phonemes **Listening:** listening for specific information
25 *What's it called in English?* Describing objects	Describing things when you don't know the word	Adjectives for shape, material, size etc. Words to describe something if you don't know the English word Everyday objects	**Listening:** listening for main ideas **Sounds:** consonant clusters; word linking in sentences **Speaking:** describing everyday objects
Progress check lessons 21 – 25	Revision	Verbs from nouns and nouns from verbs Noun suffixes Meanings of *get*	**Sounds:** /ɜː/ and /ə/; /ɔː/ and /ʌ/ **Speaking:** Talking about your travel experiences **Writing:** writing a paragraph about other students travel experiences.
26 *Safety first* Safety instructions and rules	Modal verbs *Must* for obligation; *mustn't* for prohibition	Words to describe situations where safety instructions apply: on a motorway, in a train, at the border, in the street	**Speaking:** talking about safety instructions **Sounds:** linking of /mʌst/ and /mʌsnt/; insistent intonation **Reading:** reading about safety instructions **Listening:** listening for main ideas
27 *The Skylight* A short story by Penelope Mortimer	*Can, could* (1) for ability	New words from a story *The Skylight*	**Speaking:** talking about what you can or can't do; predicting what happens next in a story **Listening:** listening for main ideas; listening for specific information
28 *Breaking the rules?* Rules and customs in everyday situations in different countries	*Can, can't* (2) for permission and prohibition	Words to describe rules in everyday situations	**Reading:** reading and answering a questionnaire **Sounds:** strong and weak forms of *can*; American English *can* and *can't* **Listening:** listening for main ideas **Speaking:** talking about rules
29 *Warning: flying is bad for your health* Advice on staying healthy	*Should* and *shouldn't* for advice	Medical complaints Parts of the body	**Reading:** reading for specific information **Listening:** listening for specific information **Speaking:** talking about advice for staying healthy
30 *Doing things the right way* Behaviour and manners in different social situations	Asking for permission Asking people to do things Offering	Words from a questionnaire about behaviour in social situations	**Reading:** reading and answering a questionnaire **Sounds:** polite intonation in questions **Listening:** listening for specific information
Progress check lessons 26 – 30	Revision	Adjectives and nouns which go together Words with more than one meaning and part of speech Techniques for dealing with words you don't understand Word association	**Sounds:** /əʊ/ and /ɔɪ/; /ʃ/, /tʃ/ and /dʒ/; polite and friendly intonation **Speaking:** talking about advice and rules for foreign visitors to your country **Writing:** writing some advice and rules for foreign visitors

Lesson	Grammar and functions	Vocabulary	Skills and sounds
31 *My strangest dream* An English woman talks about her dream in which the Queen came to tea	Past continuous (1) for interrupted actions *When*	Verbs and prepositions which go together Adjective and noun or noun and noun combinations Words for story telling	**Speaking:** predicting what happens next in a story **Listening:** listening for specific information **Writing:** writing a story using *suddenly, fortunately, unfortunately, to my surprise, finally*
32 *Time travellers* A true story about two women who travelled back in time	Past continuous (2): *while* and *when*	New words from a passage called *Time travellers*	**Reading:** predicting; reading for main ideas **Speaking:** talking about travelling in time
33 *Is there a future for us?* Two children give their views on the environment in the future	Expressions of quantity (2): *too much/many, not enough, fewer, less* and *more*	Geographical features and location	**Reading:** reading for specific information; inferring **Speaking:** talking about the geography of your country; talking about the environment
34 *The day of the dead* An article about Mexico's Day of the Dead	Present simple passive	Religion Rituals and festivals	**Reading:** predicting; reading for main ideas; reacting to a passage **Speaking:** talking about a ritual or festival in your country **Writing:** writing about a ritual or festival
35 *Mind your manners!* Table manners and behaviour in social situations	Making comparisons (3): *but, although, however*	Food Plates, cutlery etc. Cooking utensils	**Reading:** reading and answering a questionnaire **Listening:** listening for specific information **Sounds:** stress and intonation in sentences with *but, however, although* **Speaking:** talking and writing about table manners and social occasions in your country
Progress check lessons 31 – 35	Revision	Multi-part verbs	**Sounds:** /v/ and /w/; /h/; stress in multi-part verbs **Writing:** punctuating a story; inserting words into a story
36 *Lovely weather* The best times to visit different countries	*Might* and *may* for possibility	Weather	**Reading:** reading for specific information **Writing:** writing a letter giving advice about the best time to visit your country
37 *Help!* Emergency situations	First conditional	Words to describe emergency situations	**Speaking:** talking about emergency situations; predicting the end of a story **Sounds:** /l/; word linking in sentences **Listening:** listening for main ideas, listening for specific information
38 *My perfect weekend* Two people describe their perfect weekend	*Would* for imaginary situations	Luxuries and necessities New vocabulary from a passage called *My perfect weekend*	**Speaking:** talking about luxuries and necessities; talking about your perfect weekend **Reading:** reacting to a passage **Sounds:** linking of /d/ ending before verbs beginning in /t/ or /d/ **Listening:** listening for main ideas
39 *The umbrella man* A short story by Roald Dahl	Second conditional	New vocabulary from a story called *The umbrella man*	**Reading:** predicting; reading for main ideas; reading for specific information **Listening:** listening for specific information **Writing:** rewriting a story from a different point of view
40 *How unlucky can you get?* Unlucky experiences	Past perfect: *after, when* and *because*	New vocabulary from a story called *How unlucky can you get?*	**Listening:** listening for specific information **Sounds:** linking of /d/ in past perfect sentences **Speaking:** predicting the end of a story **Writing:** writing sentences using *after, because* and *when*
Progress check lessons 36 – 40	Revision	*Make* and *do* Formation of adverbs	**Sounds:** /w/, /r/; /ɔː/, /aʊ/; stressed words **Speaking:** talking about difficult situations; preparing and acting out a dialogue

Communication activities Grammar section Irregular verbs Phonetic alphabet

1 Welcome!

Present simple (1) for customs and habits: questions; adverbs of frequency

SPEAKING AND LISTENING

1 Work in pairs. Where do you hear these words and phrases?
– in a bar – in a shop – in a hotel – at home – in class

Hello. Goodbye. Come in. How do you do. Pleased to meet you.
Can I help you? How much is this? Thank you. Fine, thanks.
Sorry! I'd like a Coca Cola. I don't understand. How are you?
Could you repeat that? This is my friend, Rosario.

Which ones go together?
Hello. How are you?

2 🔲 What are the situations? Listen and find out.

3 Put the words in the right order and make questions. Then underline the stressed words.

1 first your what's name
2 are old you how
3 much earn how you do
4 do live you where
5 are married you
6 do do you what
7 and you brothers sisters do any have
8 from where come do you

What's your first name?

🔲 Now listen and check. Say the sentences aloud.

4 Look at the questions in 3. Which ones are suitable questions for:
– someone you know well   – someone you don't know well?

5 Go round the class greeting people and asking suitable questions.

READING

1 *Make yourself at home* is about hospitality in Germany, Saudi Arabia, Britain and Japan. Read it and match these headings with the paragraphs.

a type of clothes   b length of stay   c refreshments   d special customs
e gifts   f topics of conversation   g time of arrival

2 Work in pairs. Match the countries and the paragraphs.
I think paragraph 1 is Japan because they sit on the floor in Japan.

3 Which paragraphs are true for your country?

Make yourself at home

1 'In my country, men usually go to restaurants on their own. They always take their shoes off before they go in. Then they usually sit on the floor around a small, low table. In the evening they often sing songs.'

2 'You usually take chocolates or flowers. But you always take an odd number of flowers, and you remove the paper before you give them to the hostess. You can also send flowers before you arrive. You don't usually take wine except when you visit very close friends.'

3 'We always offer our guests something to drink when they arrive, tea, coffee or perhaps water or soft drinks. We think it is polite to accept a drink even if you're not thirsty. If you visit someone you always stay for a few drinks. When you have had enough to drink, you tap your cup or put your hand over it. If you say no, your host will insist that you have more to drink.'

4 'People's private lives are very important so they never ask you personal questions about your family or where you live or your job. They never talk about religion or matters of finance, education or politics, but usually stay with safe subjects like the weather, films, plays, books and restaurants.'

5 'It's difficult to know when to leave, but an evening meal usually lasts about three or four hours. When the host serves coffee, this is sometimes a sign that the evening is nearly over, but you can have as much coffee as you want.'

6 'If the invitation says eight o'clock then we arrive exactly at eight. With friends we know well, we sometimes arrive about fifteen minutes before.'

7 'Obviously it depends on the occasion, but most dinner parties are informal. The men don't usually wear a suit, but they may wear a jacket and tie. Women are usually smart but casual.'

GRAMMAR

Present simple (1) for customs and habits
You use the present simple to talk about customs and habits.
*In my country men **go** to restaurants on their own. They **take** their shoes off.*

Negatives
*He **doesn't** live here.   You **don't** take wine.   We **don't** ask personal questions.*

Questions
There are two types of questions:
– with question words: ***who, what,** etc.*
***What's** your first name?   **How are** you?*
– without a question word
***Are** you married?   **Do** you **have** any brothers and sisters?*
You can answer this type of question with *yes* or *no*.

Adverbs of frequency
You can use adverbs of frequency to say how often things happen.
*They **always** take their shoes off.   We **sometimes** arrive early.*
*We **usually** take chocolates or flowers.   We **never** ask personal questions.*
*We **often** wear jeans and sweaters.*

1 Complete these sentences with verbs from the passage.
1 In my country we ___ at a table for our meals.
2 We usually ___ ten or fifteen minutes after the time on the invitation.
3 People ___ coffee at the end of the dinner party and then they ___ .
4 Men often ___ a suit when they ___ to a restaurant.

2 Write a question for each heading in *Reading* activity 1.
a type of clothes: What do you wear to a dinner party?

3 Look at *Make yourself at home* again. Underline the verbs which are with an adverb of frequency.

WRITING AND VOCABULARY

1 Choose seven or eight verbs from the box and write sentences about hospitality in your country.

accept	arrive	ask	answer	come from	drink	earn	give	go	know	live
offer	put	say	send	sing	sit	stay	take	take off	talk about	think	visit
want	wear

We offer a cup of tea when guests arrive.

2 Now add a suitable adverb of frequency to each sentence.
We usually offer a cup of tea when guests arrive.

3 Look back at the questions you wrote in *Grammar* activity 2, and write answers for your country.

2 *A day in the life of the USA*

Present simple (2) for routines: third person singular; expressions of time

READING AND VOCABULARY

1 Read *A day in the life of the USA* and decide who you can see in the photos.

2 Work in pairs. Say what you usually do at the times mentioned in the passage.
At 6.30am I'm still asleep.
I also have lunch at 12.45pm.

3 Are there any differences from life in your country ?
In my country we usually have lunch at 3pm.

4 Match the verbs with suitable nouns in the box.

start television come dinner leave have stop lunch watch breakfast home get school finish work

start school, start work...

5 Say what time you do these things.

get up   get dressed   go shopping go to sleep   have a shower/bath wake up   wash   wash up

I get up at seven thirty.

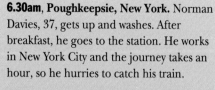

A day in the life of the USA

6.30am, **Poughkeepsie, New York.** Norman Davies, 37, gets up and washes. After breakfast, he goes to the station. He works in New York City and the journey takes an hour, so he hurries to catch his train.

7.15am, **Roanoke, Virginia.** A tired Annie Laurence, 10, wakes up and gets ready for school. An hour later she leaves home. She has lunch at school, usually sandwiches and an apple. It's a long day for Annie. She doesn't get home again until 5pm at the end of the afternoon.

10.30am, **Long Beach, California.** Tony de Valera takes a coffee break between meetings. He works for the Disney corporation as an imagineer, a job that is somewhere between an artist, an engineer and a science-fiction writer.

12.45pm, Evanston, Illinois. Thirty-four-year-old Amelia Noriega, head of public relations for a major car manufacturer, stops work and goes shopping. Then she has lunch with a friend. 'There aren't many women at my job level,' she says. 'But there are more every day.'

4.30pm, Tampa, Florida. George Markopoulos, 65, comes home after his daily swim. Then he joins his wife at the community centre, where she teaches physical education. 'I feel twenty years younger than I am.'

6.15pm, Seattle, Washington. Jo-Ann Rosenthal leaves work after a long day as a telephonist at a downtown bank. It's Friday night so she walks to her local bar and meets her friends.

7.45pm, Lubbock, Texas. Cliff Renton III, 61, meets Walter Avery, 62, to have dinner and to talk about the local Ranch Handlers' Ball, the most important event in the Lubbock social calendar. Cliff is president of the Social Committee, so he's responsible for the success of the evening.

11pm, Athens, Georgia. Shirlee Lewis finishes dinner, washes up and watches the TV news. Her five children are asleep, so she tries to be very quiet.

GRAMMAR

Present simple (2) for routines: third person singular

You use the present simple to talk about routines.

*He **gets up** at 6.30. She **works** in Seattle.*

You form the third person singular (*he/she/it*) of most verbs in the present simple by adding -s.

He gets up at 6.30. She works in Seattle.

You add -es to *do*, *go* and verbs which end in -ch, -ss, -sh and -x.

He washes. She goes to school.

Verbs which end in a consonant + -y change to -ies.

carries flies

The third person singular of *be* is *is*. You often use the contracted form *'s*.

It's a long day for Annie.

The third person singular of *have* is *has*.

She has lunch at school.

Expressions of time

***in** the morning **in** the afternoon **in** the evening*
***at** night*
***before** lunch **after** dinner at **about** seven o'clock*

---

1 Write down the third person singular of these verbs. (You can find them in the passage.)

come join finish get go hurry leave meet
say stop take teach try wake walk wash
watch work

2 Put the verbs in three columns.

-s	-es	-ies
gets	goes	hurries

Now add these verbs to the correct column.

do dress fly live make carry

3 Say what time of day you do the things in *Reading and vocabulary* activity 5. Use the expressions of time in the box above.

LISTENING

1 🔲 Listen and decide which people in *A day in the life of the USA* are speaking.

2 🔲 Listen again and find out what they do at these times:

Speaker 1: 8.00am 1.00pm 6.30pm 11.30pm
Speaker 2: 8.15am 12.30pm 5.30pm 7.00pm

3 Work in pairs and check your answers.
At 8.00am ___ leaves home and goes to work.

SOUNDS

1 There are three different ways of pronouncing the final -s in the third person singular present simple.

🔲 Listen to these verbs. Is the final sound /s/ , /z/ or /ɪz/? Put them in three columns.

takes goes finishes sits sings arrives refuses
offers has asks talks serves washes watches
does

2 Now say the words aloud.

🔲 Listen and check.

SPEAKING AND WRITING

1 Work in pairs and find out about your morning routines. What time does your partner do these things?

– wake up – get up – get dressed
– have breakfast – go to work

Cecile, what time do you wake up?
I wake up at seven o'clock.

What time do you get up?
At a quarter-past seven.

Make notes about your partner's routine.
Cecile – wake up: 7am, get up: 7.15am.

2 Write sentences about your partner's morning routine.
Cecile wakes up at seven o'clock.
She gets up at seven fifteen.

3 Now write a paragraph about your partner's morning routine. Link the sentences you wrote in 2 with *and* and *then*.
Cecile wakes up at seven o'clock and gets up at seven fifteen. Then she...

What does the word 'home' mean to you? How do you say the word in French? In Spanish? In your language? Although people usually know what the word means, it often has no exact translation. It's not surprising really, because the idea of home varies from country to country, and from person to person. A home is more than a roof and four walls. It's the cooking, eating, talking, playing and family living that go on inside which are important as well. And at home you usually feel safe and relaxed.

Home rules

But it's not just that homes look different in different countries, they also contain different things and reveal different attitudes and needs. For example, in cold northern Europe, there's <u>a</u> fire in the living room or kitchen and all the chairs face it. In the south, where <u>the</u> sun shines <u>a lot</u> and it's more important to keep the heat out, there are small windows, cool stone floors and often no carpets. We asked some people about their homes.

What's the main room in your home?
'The kitchen, because its warm and we have <u>breakfast</u>, <u>lunch and dinner</u> there seven days a week.' **Jackie, Cork , Ireland**

Do you have a television? If so, where?
'In the bedroom. We like to watch it <u>in bed</u>.'
Maurice, Bruges, Belgium

Do you lock your door when you go out?
'In cities we do. Although when I was a child in <u>the</u> Tatra mountains, we left <u>the</u> door open with bread and dishes of food and something to drink, such as a glass of milk, on a table inside, so that visitors and travellers could stop and refresh themselves.' **Grazyna, Katowice, Poland**

How often do people move home in your country?
'In the USA many people move every ten years or more.'
Cheryl, Boston, USA

If you live in a town, do you stay there at weekends?
'Well, we live in the town, but only because I'm <u>an</u> architect and I work there. I really wouldn't call it home – that's what I call our house in the country where we go every weekend.'
Elizabeth, Sao Paulo, Brazil

What are typical features of homes in your country?
'In Britain, even in the town there's always a garden and sometimes a cellar. We have <u>separate bedrooms and living rooms</u>. But we don't often have balconies or terraces. The weather isn't warm enough!' **Pat, Exeter, England**

So *home* means different things to different people. What does it mean to you?

VOCABULARY

1 Look at the words in the box and find *two types of housing* and *six rooms*.

basin bath bathroom bed bedroom carpet chair cooker cupboard curtains dining room dishwasher door flat fridge garden house kitchen lamp living room shower sink sofa table toilet video washing machine window

2 Work in pairs. In which rooms do you find the furniture and equipment in the box above? Which word is left?

3 Work in pairs. Ask and say what rooms there are in your homes.

Is there a living room?
Yes, there is.
Is there a dining room?
No, there isn't.

Now ask and say what furniture and household equipment there is in your home and where it is. Add words to the lists if you can.

There's a table and some chairs
in the kitchen.
Is there a video? No, there isn't.

READING

1 Read the first two paragraphs of *Home rules*. What does *home* mean to the writer? Do you agree?

2 Read the rest of the passage and think about answers to the questions for your country.

GRAMMAR

> ### Articles
> **You use the indefinite article *a/an*:**
> – **to talk about something for the first time:** *There's **a** kitchen and **a** dining room.*
> – **with jobs:** *I'm **a** teacher.   She's **an** engineer.*
> – **with certain expressions of quantity:** ***a** little food, **a** few beds, **a** couple of friends*
>
> **You use the definite article *the*:**
> – **to talk about something again:** *In **the** kitchen there's a table, and on **the** table there's a cat.*
> – **with certain places and place names:** ***The** Alps, **The** West, **The** USA*
> – **when there is only one:** ***the** president, **the** government, **the** weather*
>
> **You don't use an article:**
> – **with plural and uncountable nouns when you talk about things in general:** *It's got carpets and curtains.   There's lots of food.*
> – **with certain expressions:** *at home, at work, in bed, by car*
> – **with meals, languages, most countries and most towns:** *Let's have lunch.   Speak English.   We live in France.   I lived in Paris.*
>
> ### Plurals
> **You form the plurals of most nouns with *-s*:** *chair – chairs   cupboard – cupboards*
> **You add *-ies* to nouns of two or more syllables which end in *-y*:** *balcony – balconies*
> **You add *-es* to nouns which end in *-ch*, *-ss*, *-sh*, and *-x*:** *church – churches*
> **There are some irregular plurals:** *man – men   woman – women   child – children*

1 Look at the nine articles and phrases underlined in the passage. Which of the rules about articles in the grammar box do they illustrate?

2 Complete the sentences with *a/an*, *the* or put – if there's no article.

1 Last year we moved to ___ London.
2 ___ kitchen is ___ door on your left.
3 ___ weather is very hot in August.
4 There isn't ___ table in ___ kitchen.
5 Would you like ___ drink?
6 I'm sorry, he's still at ___ work.

3 Write the plural of these nouns.

parent house city family dish party bush country table fax feature

SPEAKING

1 Work in pairs and look at the photo. What room do you think it is? Does it look like a room in a house in your country?

2 Work in pairs and talk about your answers to the questions in *Home rules*.

3 What does the word 'home' mean to you? Write five words or phrases which you associate with the idea. Find out what other students in your class wrote.

Verb patterns (1): *-ing* form verbs; talking about likes and dislikes

First impressions

The British and the Americans speak the same language. But life in the two nations can be very different. We asked some Americans what they like or don't like about Britain...

'The police. They're very friendly and they don't carry guns.' Claude, Trenton

'The weather is awful. You don't seem to get any summer here. It's winter all year round.' Toni, San Francisco

'The tourists! The streets are so crowded. I think you should do something about them. And I can't stand the litter everywhere. It's a very dirty place.' José, Washington

'Walking and sitting on the grass in the parks, especially on a hot summer's day. Oh, and the green countryside. But why is the beer warm?' Max, Houston

'Well, they certainly seem rather unfriendly. Nobody ever talks on the buses. But maybe we haven't met any real English people yet.' Eva, Niagara Falls

'Feeling safe when you walk the streets. Oh, and the polite drivers who stop at a street crossing if they see someone waiting there.' Moon, Los Angeles

'Driving on the left. It's very confusing. I keep looking the wrong way.' Paula, San Diego

So then we asked some British people what they like or don't like about America...

'Arriving at the airport. Immigration is so slow, it takes hours to get through!' Geoff, London

'The waste of electricity. I just can't understand why their homes are extremely hot in winter and very cold in summer.' Louise, Southampton

'The people, they're so generous. If they invite you home, you're sure of a big welcome.' Amin, Bath

'Going shopping. I love it. It's so cheap everywhere – food, clothes, hotels, petrol.' Paul, Oxford

'I hate the insects. They're so big. In Texas the mosquitoes are enormous. But I suppose in Texas they would be!' Maria, Glasgow

'Lying on the beach in the sunshine. In California the sun shines all day, every day. It's great.' Rose, Cardiff

'Driving on the right. It's very confusing. I keep looking the wrong way.' Paula, St Albans

READING AND VOCABULARY

1 Read *First impressions,* which is about people's impressions of Britain and America. Put a tick (✔) by the positive impressions and a cross (✘) by the negative impressions. Which is the most surprising impression?

2 Underline the adjectives in the box.

> awful beach cheap cold confusing countryside crowded dirty driving
> food friendly generous grass great gun hot insect park police polite
> shopping slow summer sunshine tourist unfriendly walking weather winter

Now write them in two columns, *positive* and *negative,* according to their meaning in the passage.

Can you remember which nouns they went with in the passage?

3 Look at the other words in the box. Put a tick (✔) by anything you particularly like and a cross (✘) by anything you dislike.

Think of two or three things that people like and dislike about your country.

GRAMMAR AND FUNCTIONS

Verb patterns (1): *-ing* form verbs
You can put an *-ing* form verb after certain verbs.
*I **love** walking. She **likes** swimming. They **hate** lying on a beach.*

Talking about likes and dislikes

Questions	**Short answers**
***Do** you **like** the weather?*	*Yes, I **do**.*
***Does** he **like** the police?*	*Yes, he **does**.*
***Do** you **like** walking in the park?*	***No, I** don't.*
***Does** she **like** the weather?*	***No, she** doesn't.*

Negatives
*I **don't** like the weather. He **doesn't** like arriving at the airport.*
*We **don't** like insects. She **doesn't** like driving on the right.*

Expressing likes	**Expressing dislikes**
I love it.	*I hate it. I can't stand it.*
A lot.	*Not at all. I don't like it at all.*
A little.	*Not very much.*

Expressing neutrality
It's all right. I don't mind.

Expressing the same likes and dislikes
I like rock music. So do I.
I don't like jazz. Neither/Nor do I.

Expressing different likes and dislikes
I like rock music. I don't.
I don't like jazz. I do.

1 Do you like or dislike these things? Respond to the statements.

1 I like going to parties.
2 I don't like cooking.
3 They like walking.
4 I don't like insects.
5 He doesn't like tennis.
6 She likes shopping.

1 I don't.

2 Look back at the passage and say what the people like or dislike and why.
Claude likes the British police because they're friendly and don't carry guns.

SOUNDS

Listen to the way these people use a strong intonation to express strong likes or dislikes.

1 I hate the insects.
2 I can't stand the litter.
3 He loves shopping.
4 I like walking very much.
5 She doesn't like the weather at all.
6 He hates the warm beer.

Now say the sentences aloud.

SPEAKING

1 Write down four or five things you like about your town or country, or the town where you are now.

2 Find people in the class who like the same things. Now talk about things you don't like.

3 Tell the rest of the class about your likes and dislikes. Make two class lists: the *Top five likes* and the *Top five dislikes* about your town or country, or the town where you are now.

5 *Take a closer look*

Present simple and present continuous

SPEAKING

1 Work in pairs. Take a closer look at your partner! How much do you know about each other? Guess the answers to these questions.

Does he or she...

– play a musical instrument?
– paint or draw?
– travel by bus often?
– smoke?
– work hard?
– speak any foreign languages?
– listen to music?
– earn a lot of money?

I think she plays the piano.

2 Now ask and answer the questions about each other. Did you guess correctly in 1?
Do you play a musical instrument?   No, I don't.

3 Look at the *yes* answers. Is your partner doing these things at the moment?
Is he working hard?   Yes, he is.
Is she smoking?   No, she isn't.

4 Look at the photos. Choose one person in each photo and take a closer look. Imagine what their life is like and guess the answers to the questions in 1.
I think he plays a musical instrument.

GRAMMAR

Present simple

You use the present simple to talk about:
- **a habit**   *He smokes twenty cigarettes a day.*
- **a personal characteristic**   *She plays the piano.*
- **a general truth**   *You change money in a bank.*

There is an idea that the action or state is permanent.

Present continuous

You use the present continuous to say what is happening now or around now.

It's raining.   He's drawing a picture.   I'm learning English

There is an idea that the action or state is temporary.

You form the present continuous with *is/are* + present participle (verb + *-ing*).

*I'm look**ing** at the photos.   She**'s** wait**ing** for a bus.*

Questions	**Short answers**	**Negatives**
*Is he draw**ing?***	*Yes, he **is**. No, he **isn't**.*	*He **isn't** drawing.*
*Are you go**ing** home?*	*Yes, I **am**. No, I'm **not**.*	*I'm **not** go**ing** home.*

You don't usually use these verbs in the continuous tenses:

believe feel hear know like see smell sound taste think understand want

---

1 Work in pairs and point at the people in the photos. Ask and say what they're doing at the moment and why.

What's the man in front doing?   He's playing the accordion.
Why is he playing the accordion?   Maybe he doesn't have any money.

Use these words and phrases to help you.

cross the street   draw   do the shopping   go home   hold an umbrella
listen to music   paint a picture   play the accordion   rain
shelter his instrument   sit on a suitcase   stand by the road   talk to someone
wait for a bus

2 Look at these verbs in the present continuous and write their infinitives.

drawing   getting   having   making   playing   shopping   putting   staying

draw   get ...

What happens to infinitives ending in *-e, -t, -p* and *-y* when you form their present participle?

3 Complete these sentences with *and* or *but*.

1 I often go shopping at the supermarket ____ I'm going there now.
2 They usually eat at home ____ today they're having dinner in a restaurant.
3 She walks to work ____ this week she's taking the bus.
4 He smokes ten cigarettes a day ____ he's smoking a cigar at the moment.

SOUNDS

🔲 Listen and tick the phrase you hear. Is the underlined sound /n/ or /ŋ/?

1 carry i<u>n</u>/carry<u>ing</u> an umbrella
2 sitti<u>ng</u>/sit i<u>n</u> there
3 sing i<u>n</u>/sing<u>ing</u> tune
4 arrive i<u>n</u>/arriv<u>ing</u> time
5 take i<u>n</u>/tak<u>ing</u> money
6 stand i<u>n</u>/stand<u>ing</u> there

Now say the phrases aloud.

LISTENING AND VOCABULARY

1 Say where you do these things.

buy train tickets   have dinner
change money
get some medicine   buy food

Choose from these places.

bank   chemist   post office
supermarket   railway station
restaurant

2 🔲 Listen to four conversations and decide where the people are. Choose from the places in 1. Now work in pairs. Say where the people are and what they're doing.

3 Look at the words in the box. Are they nouns, verbs or both?

---

bank bus buy change chemist
close cross draw food get
hold look at medicine money
paint play post office put
queue railway station rain road
shelter shop sit stand stay
street suitcase supermarket take
think ticket town umbrella
wait for   walk

---

4 Group any words which go together.

bank, money...

Progress check 1-5

VOCABULARY

1 Look at this crossword.

	P	O	L	I	T	E		
		O		O			A	
T	E	L	E	V	I	S	I	O	N
		I		E			S	
		C					W	
	W	E	A	R			E	
							R	

Work in pairs. Choose words in the vocabulary boxes from lessons 1 – 5 and put them in a crossword. How many words can you find?

2 Some words can be more than one part of speech. For example:
*cook: A **cook** (noun) is someone who **cooks** (verb) food.*
*orange: An **orange** (noun) is an **orange** (adjective) fruit.*

Use your dictionary to find out what parts of speech these words can be.

talk   head   drink   flat   start
rent   slice   heat

Write sentences showing their different parts of speech.
He talks all the time.
There's a talk on insects tonight.

3 Not every new word is useful to you. Look at the vocabulary boxes in lessons 1 – 5 again and choose ten words which are useful to you.

Start a *Wordbank* in your Practice Book for useful words and phrases. Write the ten words in your *Wordbank*.

GRAMMAR

1 You meet Tanya, from Russia, at a party in London. Here is some information about her. What are the questions?

1 I live in Moscow.
2 Yes, I am and we have three children.
3 I'm a scientist.
4 I start work at eight in the morning.
5 I finish at six in the evening.
6 In the evenings we have dinner.
7 At weekends we go to the country.
8 We usually go to a beach on the Black Sea.

1 Where do you live?

2 Put an adverb of frequency in each sentence so that it is true for you or your country.

always   (not) usually   (not) often   sometimes   never

1 We take off our shoes before we go into a house.
2 We offer guests something to eat.
3 We talk about politics and our families.
4 We have dinner at seven o'clock.
5 We give the hosts some wine or flowers.
6 We sit on the floor.
7 We wear smart clothes.
8 We arrive ten or fifteen minutes late.

3 Answer the questions with one of these expressions:

Yes, I do.   No, I don't.   I love it.   I hate it.
Not very much.   It's all right.   Not at all.

1 Do you like living in a town?
2 Do you like shopping?
3 Do you like walking?
4 Do you like hot weather?
5 Do you like warm beer?
6 Do you like fast food?

4 Write the *-ing* form of these verbs.

close   visit   fly   wear   like   sing   get   throw   cross   go   stay   cut

5 Choose the correct verb form.

1 Tanya *comes/is coming* from Russia.
2 She *visits/is visiting* London.
3 She *has/is having* a holiday.
4 She *speaks/is speaking* English quite well.
5 She *stays/is staying* with friends in London.
6 She *enjoys/is enjoying* her visit.
7 She *goes/is going* shopping most days.
8 She *says/is saying* she wants to come back soon.

6 Complete these sentences with *a/an, the*, or put – if there's no article.

1 There's ___ radio in ___ living room.
2 Would you like ___ cup of ___ tea?
3 They've got ___ large house in ___ centre of town.
4 We've got ___ son and ___ two daughters.
5 ___ children are outside in ___ garden.
6 What's ___ main room in your flat?
7 She spoke ___ very good French at ___ home.
8 I flew to ___ Lyon and spent ___ two weeks in ___ Alps.

SOUNDS

1 Say these words aloud.

does   finishes   gets   goes   leaves   looks   refuses
sings   smokes   wants

Do they end in /s/ , /z/ or /ɪz/? Put them in three columns.

🔲 **Now listen and check.**

2 🔲 Listen and say these words aloud.

<u>ea</u>t   <u>i</u>t   fift<u>ee</u>n   f<u>i</u>fty   l<u>i</u>ve   l<u>ea</u>ve   s<u>i</u>t   s<u>ea</u>t

Is the underlined sound /ɪ/ or /iː/? Put the words in two columns.

3 🔲 Listen to these questions. Put a tick (✓) if you think the speaker sounds interested.

1 What's your name?
2 How old are you?
3 Where do you live?
4 Do you live with your parents?
5 Are you married?
6 Is that your brother?
7 Is that your husband?
8 Do you have a sister?

Now say the sentences aloud. Try to sound interested.

SPEAKING AND WRITING

1 Look at some rules for punctuation in English.

1 You use a capital letter:
  – at the beginning of a sentence: *I like cakes.*
  – for names: *Sue   Jim   Fiona*
  – for the first person singular pronoun: *I*
  – for nationalities: *She's British   the French*
2 You put a full stop at the end of a sentence: *I don't speak Japanese.*
3 You put a question mark at the end of a question: *Do you speak English?*
4 You put inverted commas and a comma around someone's actual words: *'I'm French,' he said.*
5 You put an apostrophe for contractions: *I'm English.   He doesn't live here.*

Now punctuate these sentences.

1 we dont usually visit people without an invitation
2 when we meet people for the first time we say how do you do
3 when do you use first names in your country
4 your friend is called james smith do you call him james or mr smith
5 its usual to use first names with people when you get to know them

2 Work in groups of three or four. Talk about customs and traditions of hospitality in your country. Talk about the following:

Invitations	Do you ever visit people without an invitation?
Greetings	How do you greet people when you meet them for the first time?
Names	When do you use first names, family names or titles?
Punctuality	When do you arrive for an appointment?
Clothing	What do you wear for a lunch or a dinner engagement?
Food and drink	What food or drink do you expect?
Conversation	What do you talk about? What don't you talk about?
Compliments	Do you make compliments about the food, the host's house or personal objects?
Leaving	When do you leave a dinner party?

3 Write some advice for visitors to your country about customs and traditions of hospitality and entertainment. Write about the points in 2 and say what people do or don't do.

Surprising behaviour

Past simple (1): regular and irregular verbs

VOCABULARY AND READING

1 Look at the photo, which shows a scene from the passage you are going to read. Where do you think the passage takes place?

2 Look at the words in the box. Decide which are nouns and which are adjectives. Look up words you don't know in the dictionary.

> bag smile broken cloud
> bunch coat people flat flower
> heavy low scarf shoes big
> sky warm

3 Group the words which can go together.

heavy bag, heavy cloud…

4 The passage is by Paul Theroux, an American travel writer. Read it and find out what country he is describing.

5 Are these statements about the passage true or false?

1 The woman had a friend with her.
2 She didn't have a dog.
3 The weather wasn't very good.
4 He didn't know her name.
5 He expected her to greet him.
6 She didn't say 'Good morning'.

Which statement describes what surprised him?

---

*A*s soon as I left Deal, I saw a low flat cloud, iron-grey and then blue across the Channel. The closer I got to Dover, the more clearly it was defined. I walked on and saw it was a series of headlands.
It was France.

Ahead on the path was a person, down a hill four hundred yards away; but whether it was a man or a woman I could not tell. Some minutes later I saw her scarf and her skirt, and for more minutes on those long slopes we walked toward each other under the big sky. We were the only people visible in the landscape – there was no one behind either of us. She was a real walker – arms swinging, flat shoes, no dog, no map. It was lovely, too: blue sky above, the sun in the southeast, and a cloudburst hanging like a broken bag in the west. I watched this woman, this fairly old woman, in her warm scarf and heavy coat, a bunch of flowers in her hand – I watched her come on, and thought I am not going to say hello until she does.

She did not look at me. She drew level and didn't notice me. There was no other human-being in sight on the coast, only a fishing boat. Hetta Poumphrey – I imagined that was the woman's name – walked past me, and still stony-faced.

'Morning!' I said.

'Oh.' She turned her head to me. 'Good morning!'

She gave me a good smile, because I had spoken first. But if I hadn't, we would have passed each other, Hetta and I, in that clifftop meadow – not another soul around – five feet apart without a word.

Adapted from *The Kingdom by the Sea,* by Paul Theroux

GRAMMAR

> ### Past simple (1): regular and irregular verbs
>
> You use the past simple to talk about a past action or event that is finished.
> *We **walked** toward each other.*
> *She **turned** her head to me.*
>
> **You form the past simple of most regular verbs by adding -ed.**
> *walk – walk**ed***   *I walked on.*
> *watch – watch**ed***   *I watched this woman.*
>
> **Many verbs have an irregular past simple form.**
> *leave – left   see – saw   be – was/were*
>
> **For a full list of irregular verbs, see Grammar review.**
>
> **Negatives**
> *She **did not** look at me.   She **didn't** notice me.*

1 Look at the passage again. Find the past simple of these verbs.

be get give imagine leave see think walk watch

Which verbs are regular? Which ones are irregular?

2 Look at the past simple form of these regular verbs and write their infinitive form.

carried closed continued danced decided liked lived stopped travelled tried

What's the rule for forming the past simple of regular verbs ending in *-e, -y, -p, -l*?

3 Look at the past simple form of these irregular verbs and write their infinitive form.

became came chose cost cut did hit had heard knew made met put ran read said shut took told understood went wrote

4 Which verbs in 3 have the same form in the present and the past simple? Which verb has the same form but sounds different?

SOUNDS

1 🔲 Listen to the pronunciation of the past-tense endings of these verbs.

/t/	/d/	/ɪd/
liked	*lived*	*decided*
washed	*stayed*	*visited*

Put these verbs in the correct column.

continued finished enjoyed started walked danced called wanted expected

2 🔲 Listen and check. What's the rule? Now say the verbs aloud.

SPEAKING

1 Work in pairs. Why didn't the woman in the passage say hello to the writer first? Why did he find this behaviour surprising?
Perhaps she was...   Maybe she didn't...

2 Talk about a situation at home or in a foreign country when you found someone's behaviour surprising.
In England no one talks on the buses and underground. In my country we talk all the time!

Past simple (2): questions and short answers

When was the last time you had a holiday? And did you organise the trip or did you take a package tour? These days, most people choose a package tour, especially when they go abroad on holiday. They pay for their travel and accommodation in their own country, and they take traveller's cheques which they exchange for local money when they arrive in the foreign country. But in the past it was very different. In fact, before the middle of the nineteenth century, travelling for pleasure was rare and very expensive, and only a few rich people travelled abroad. The man who changed all this and brought in the age of mass tourism was Thomas Cook.

Thomas Cook was a printer in Leicester, England and the secretary of a local church organisation. In 1841 it was his job to arrange rail travel for members of his church to a meeting in Loughborough, a round trip of twenty-two miles. This was the world's first package trip. After this first success, he organised many more for his church. Then in 1845 he advertised a package tour to Liverpool for the general public, and before it took place he went to Liverpool to meet the hotel staff, and check the accommodation and restaurants.

He then started to organise trips all over Britain, including the Great Exhibition in London. In 1851 he published the world's first travel magazine which had details of trips, advice to travellers and articles and reports about the places to visit.

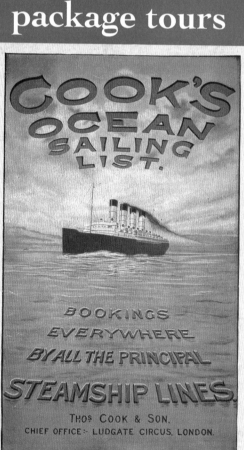

In 1854 he gave up his job as a printer. In 1855 he took his first group of tourists to Paris and later that year led a tour of Belgium, Germany and France. In 1863 he went to Switzerland and in 1864 to Italy. By then he had a million clients. The following year he opened an office in London, which his son John Mason managed. They introduced a circular ticket which gave the traveller a single ticket to cover one journey instead of a number of tickets from all the railway companies involved, and they organised a system of coupons which people bought at home and exchanged in the foreign country for a hotel room and meals.

In 1866 the first group of European tourists visited New York and the Civil War battlefields of Virginia. In 1868 the Cooks went to the Holy Land with tents because there were no hotels there at that time. After the Suez canal opened in 1869, Cook created his own fleet of luxury boats to travel up the river Nile.

It was dangerous to carry large amounts of cash, so in 1874 Cook introduced an early form of traveller's cheque, which travellers could cash at a number of hotels and banks around the world.

Thomas Cook died in 1892 at the age of 84, and his son John Mason seven years later. But the age of the package tour and mass tourism was born.

READING

1 What do these words and phrases mean?

mass tourism   package tour   traveller's cheque

Now read *The world's first package tours* **and find out what the words and phrases have to do with Thomas Cook.**

2 Read these statements about the passage. Are they true or false?

1 Travel was very expensive before the middle of the nineteenth century.
2 Thomas Cook was a travel agent in 1841.
3 The first public trip was in 1845.
4 The Great Exhibition was in Liverpool.
5 The first trip outside Britain went to France.
6 In 1866 he opened an office in London.
7 Some tourists wanted to visit the Civil War battlefields.
8 The Cooks stayed in hotels in the Holy Land.
9 The traveller's cheque allowed people to travel without large amounts of cash.
10 Thomas Cook died in 1891.

GRAMMAR

Past simple (2): questions and short answers

Questions	Short answers
Was the first package trip in 1841?	*Yes, it **was**. No, it **wasn't**.*
***Did** Cook **live** in Leicester?*	*Yes, he **did**. No, he **didn't**.*

You can use *who*, *what* or *which* to ask about the subject of the sentence. You don't use *did*.
*Who **organised** the first package trip?   Thomas Cook.*

You can use *who*, *what* or *which* and other question words to ask about the object of the sentence. You use *did*.
*Who **did** he **take** on the first package trip?   500 workers.*

Compare:
Subject   *Who **introduced** traveller's cheques?   Thomas Cook.*
Object   *What **did** Thomas Cook introduce?   Traveller's cheques.*

1 Work in pairs and check your answers to *Reading* activity 2.

1 Was travel very expensive before the middle of the nineteenth century? Yes, it was.

2 Here are some answers about the passage. Write suitable questions.

1 A few rich people.
2 He was a printer.
3 Hotel staff.
4 The world's first travel magazine.
5 In 1854.
6 New York, and the battlefields of Virginia.
7 Because it was dangerous to carry lots of cash.
8 In 1899.

1 Who travelled abroad before the middle of the nineteenth century?

VOCABULARY AND SPEAKING

1 Underline the verbs from this lesson in the box.

abroad advertise allow boat
bring buy carry cash change
check choose country create
die exchange foreign give up
holiday hotel introduce lead
leave luxury meet open
organise pay publish travel
traveller's cheque want

What is the past simple tense of the verbs?

2 Look at the other words in the box. Try to remember which nouns, verbs or adjectives they go with in the passage.

Now look back and check.

3 Work in pairs. Choose the two most important events in Thomas Cook's career.
I think one was when he gave up his job as a printer.

4 Work in pairs. When was the last time you were a tourist? Where did you go? What did you do? Did you take a package tour?

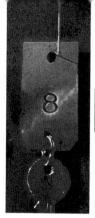

Something went wrong

Expressions of past time; *so, because*

VOCABULARY

1 Work in pairs. Put the words and phrases in the vocabulary box with the following travel situations. Some words can go with more than one situation.

a train journey   a boat journey   a plane flight
hotel accommodation

> airport   bed and breakfast   boarding pass   book
> business class   cabin   check in   check out   connection
> delay   departure   double room   fare   ferry   harbour
> land   lift   luggage   passenger   platform   reservation
> return   single   single room   terminal   ticket   take off
> timetable

2 Think of two or three more words which go with each situation.

LISTENING AND SPEAKING

1 You are going to hear Ann, an English woman, telling a story about a travel situation when things went wrong. Look at how the story begins.

> A few years ago, I was going from London to Paris to join my husband and children. I checked in early and...

What do you think the travel situation is? Which words in the vocabulary box do you expect to hear?

2 🔲 Listen to Ann's story. As you listen, look at the vocabulary box again and tick (✓) the words you hear.

3 Work in pairs. Did you guess correctly in 1? Describe what happened.

🔲 Listen again and check.

4 Look at these phrases taken from another story about a situation where something went wrong. Decide which travel situation the speaker, Bob, is describing. What do you think happened?

sat down on the tiny single bed ☐
knocked on the door ☐
slept in the car that night ☐
didn't have a reservation ☐
wanted to check out ☐
asked if he had a room ☐
left my suitcase in my car ☐
picked up a key from behind the desk ☐
was frightened by the man downstairs ☐
showed me a very dusty room ☐

5 🔲 Listen to Bob's story. Did you guess correctly in 4?

6 Number the phrases in 4 in the order you heard them.

🔲 Now listen to Bob's story again and check.

7 Have you ever been in a situation where something went wrong? Tell the class about it.

GRAMMAR

> **Expressions of past time**
> **You can use these expressions of past time to say when something happened.**
>
> *last night   last Sunday   last week   last month   last year*
> *yesterday   yesterday morning   yesterday afternoon*
> *the day before yesterday*
> *two days ago   three weeks ago   years ago   ages ago*
> *in 1985   from 1987 to 1993*
>
> **So, because**
> **You can join two sentences with *so* to describe a consequence.**
> *She often took the plane, **so** she didn't look at the safety instructions.*
> **You can join the same two sentences with *because* to describe a reason.**
> *She didn't look at the safety instructions **because** she often took the plane.*

1 Complete the sentences with these verbs in the past simple. There are some extra verbs, and there may be more than one possible answer.

ask   enjoy   finish   listen   look   open   play   start
stay   walk   visit   wash   watch

1 I _____ television last night.
2 She _____ to the news on the radio yesterday.
3 He _____ football the day before yesterday.
4 We _____ my parents last Sunday.
5 I _____ at home last weekend.

2 Look at the sentences in 1 and underline the expressions of past time.

3 Write sentences saying what you did, using the expressions of past time you underlined in 2.

4 Think of four or five important or memorable events in your life. Write down when they happened. Don't write down what happened.
five years ago   last year   three months ago...

5 Work in pairs. Show each other what you wrote in 4. Ask and say what happened.
What happened five years ago?   I went to America.

6 Join the two parts of the sentence with *because.*

1 The flight was on time
2 They had to make an emergency landing
3 He didn't have a hotel reservation
4 He decided to leave

a there wasn't much air traffic.
b the room was dirty and unpleasant.
c he didn't expect to stay there long.
d they thought there was a bomb on the plane.

1 The flight was on time because there wasn't much air traffic.

7 Rewrite the sentences in 6 using *so.*

1 There wasn't much air traffic so the flight was on time.

WRITING

1 Write an opening sentence about what happened to Bob.
Bob arrived in a town at ten o'clock.

2 When you are ready, give your sentence to another student and you will receive an opening sentence from someone else. Read it, then write another sentence to continue the story.
Bob arrived in a town at ten o'clock.
He looked for a hotel.
He didn't have a reservation.
The main hotel was full.
He went down the road to a small guest house.
The lights weren't on.
He walked past it.

Continue writing and receiving sentences until the story is finished.

3 Now write the story in full by joining the sentences with *and, but, so* and *because.*
Bob arrived in a town at ten o'clock and looked for a hotel. But he didn't have a reservation and the main hotel was full. So he went down the road to a small guest house. Because the lights weren't on, he walked past it...

4 Write Ann's story in full.

9 *Family life*

Possessive 's; possessive adjectives

VOCABULARY AND LISTENING

1 Look at the words in the vocabulary box. Put the words in pairs. Two words have no pairs. Which ones are they?

aunt boy boyfriend brother child cousin daughter father friend girl girlfriend grandfather grandmother husband man mother nephew niece parent sister son uncle wife woman	

aunt – uncle...

2 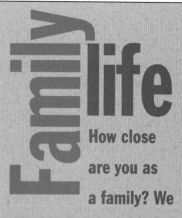 Listen to Corinne, who is French, talking about her family. Draw a line to show each person's relationship to her.

Jacqueline Raymond Chantal Christine Tony Georges Vincent Marie

mother sister father grandmother brother aunt husband uncle

GRAMMAR

Possessive 's

You add 's to singular nouns to show possession.
Corinne's father = her father   *Vincent's wife = his wife*
You add s' to regular plural nouns.
the parents' house = their house   *the boys' mother = their mother*
You add 's to irregular plural nouns.
the children's aunt = their aunt   *the men's room = their room*

Possessive adjectives

I – my you – your he – his she – her it – its we – our they – their
*Vincent is **my** husband.*

1 Work in pairs. Use the information in *Vocabulary and listening* activity 2 and say who these people are. Use the possessive 's.

1 Corinne (Raymond)   4 Georges (Marie)
2 Raymond (Chantal)   5 Marie (Corinne)
3 Chantal (George)   6 Corinne (Vincent)

1 Corinne is Raymond's sister.

2 Rewrite the sentences in 1 using possessive adjectives.
1 Corinne is his sister.

Family life

How close are you as a family? We talked to Corinne Mathieu, from Montpellier, France about her family life.

1 'We usually see each other at least once a month, maybe more often. We have lunch together on Sunday if we haven't got anything special to do. We live in Montpellier, which is about an hour and a half away, but we always come to Marseilles where my mother and father live. It's not so far. Usually my grandmother and my uncle and aunt are there too – we're quite a large family! Sometimes my brother and his girlfriend come over – they live nearby. The meal takes about four hours, we spend a lot of time chatting and there's always lots to eat.

2 'There's no one we call the head of the family, although my father's advice and opinion are very important in any decisions we take. My uncle Tony is in fact older than my father, so I suppose he's the real head of the family. When my grandfather was alive he liked to think that the whole family organised itself around him, but these days it's different. But we all try to discuss things together when we meet.

SOUNDS

1 [cassette] Listen and underline the /ə/ sound.

brother daughter father husband mother
parent sister woman

Now say the words aloud.

2 Look at this true sentence.
Vincent is Corinne's husband.

[cassette] Listen and correct the statements below with the true sentence. Change the stressed word each time.

1 Vincent is Corinne's uncle.
2 Tony is Corinne's husband.
3 Vincent is Marie's husband.
4 Vincent is Corinne's brother.
5 Vincent is Chantal's husband.
6 Georges is Corinne's husband.

*1 No, Vincent is Corinne's **husband.***

READING AND SPEAKING

1 Read *Family life* and match the questions with each paragraph. There is one extra question.

a Who's the head of the family?
b How often does the family get together?
c How many people live in your house?
d How long do people live with their parents?
e How many people live in the same house?

2 Which paragraphs give specific information about Corinne's family? Which paragraphs give more general information?

3 Work in pairs. In your country, do you talk about your family to people you don't know? If so, answer the questions above with specific information about your family. If not, answer the questions above with general information about family life in your country.

3 'In most families, it's a small family group who live in the same house, mother, father and the children before they get married. But if one of the grandparents dies, the other usually sells their home and goes to live with their children. So it's quite common to have one grandparent living with you, but not more.

4 'In France most children leave home when they get married, and not before. I lived in Marseilles with my mother and father until I got married. But there are some people who want to lead independent lives and they find a flat as soon as they start their first job, even before they get married. Of course, the main problem is that flats are so expensive to rent here, and we simply have to live with our parents.'

10 | *The town where I live*

Have got

VOCABULARY AND LISTENING

1 Look at the words in the box. Tick (✓) any features or facilities your town has got.

> art gallery beach cathedral
> cafe sports stadium museum
> nightclub park swimming pool
> university port cinema theatre
> factory restaurant

2  Listen to Catherine, an English woman who lives in Mestre near Venice. Tick (✓) the features or facilities in the box which she mentions.

3 Write down what Catherine considers to be the advantages of living in Mestre.

Listen again and check.

4 Work in pairs. Underline the adjectives in the box.

> architecture bad beautiful
> boring busy cheap climate
> cold crowded dangerous dirty
> entertainment excellent
> expensive food good
> interesting large medium-sized
> modern old safe safety
> shops size small traffic

Which adjectives can you use to describe the features and facilities of your town? Can you think of other adjectives to describe them?

5 Listen to six people talking about their home towns. Each one is talking about a different aspect of their town. Put the number of the speaker by the aspect they are talking about.

architecture	☐	climate	☐	cost of living	☐
entertainment	☐	food	☐	public transport	☐
safety	☐	traffic	☐	size	☐

GRAMMAR

> ### Have got
>
> You use **have** or **have got** when you talk about facilities, possession or relationships. You can use the contracted form with **have got**, but this is not common with **have**.
>
> *I **have** a new car.*   or   *I've got/have got a new car.*
> *My town **has** a museum.*   or   *My town's got/has got a museum.*
>
> In spoken English it is much more common to use the contracted form if possible.
>
> **Negative**
> *I **don't have** any cigarettes.*   or   *I **haven't got** any cigarettes.*
> You don't usually use **have got** in the past.
> When you talk about facilities, you can also use **there is/are**.
> *It's got an art gallery.*   or   *There's an art gallery.*

3 Join two advantages or two disadvantages together using *and*.
It's quite a small city and it's full of history.

Then join an advantage and a disadvantage together with *but*. Make sure they are all features which can go together.
*There are no cars and lorries **but** there are lots of tourists.*

4 Make a list of the advantages and disadvantages of living in your town.

Advantages	*Disadvantages*
good public transport	*too much traffic*
excellent shops	*very expensive*

5 Write sentences joining advantages or disadvantages using *and*. Then write sentences joining an advantage with a disadvantage using *but*.
It's got good public transport and the shops are excellent.
There's too much traffic but the public transport is good.

6 Write a short description of your town. Use Anna's description and the sentences you wrote in 5 to help you.

1 Rewrite these sentences with contractions.

1 The university has got a park.
2 It has got a modern tram system.
3 He has got a swimming pool.
4 I have got tickets for the theatre.
5 Rio has got some beautiful beaches.
6 She has got a good view from the window.

2 Work in pairs. Say what features and facilities your town has and hasn't got.
It's got a cathedral, but it hasn't got a theatre.

WRITING

1 Read this description of Venice by Anna, a student. Which aspects in *Listening* activity 5 does she mention?

2 Read the passage again and write down all the advantages and disadvantages of living in Venice.
Advantages: small city, full of history...
Disadvantages: lots of tourists, expensive...

The town where I live

I love living in Venice, it's full of history. I like it because it's quite a small city. I think it's got a population of about 200,000 people, but there are lots of tourists, especially in the summer, and there's not enough room for them all. Of course, there are no cars or lorries, but there are water buses on the main canals all day and night, and it's not far to walk anywhere in the city. But you need lots of money to live here. Flats are very expensive and everything comes from the mainland, so the prices of everyday supermarket items are rather high. And apart from cinemas and theatres, there's not much to do in the evenings. But I still love it here.

Progress check 6-10

VOCABULARY

1 Work in pairs and make a *word chain*. Each word must be associated with the word immediately before it. Start like this:

hotel   reception   reservation   double room
bed   breakfast

2 Some nouns go together to make a new word. Sometimes you write them as two words: *double room*, and sometimes as one word: *timetable*.

Put a noun from list A with a noun from list B and make at least 12 new words.

A package   travel   round   return   train   first
booking   boarding   air   suit

B journey   pass   terminal   class   case   tour   trip
agent   ticket   office

3 It is useful to write down new vocabulary under headings, such as *travel, towns, family* etc. You can also use more personal or impressionistic categories, such as words that sound nice, or words that look like words in your own language. Look at the vocabulary boxes in lessons 6 – 10 again. Choose words which are useful to you and group them under headings of your choice in your *Wordbank*.

GRAMMAR

1 Write these regular verbs in the past simple tense.

ask   carry   change   continue   decide   enjoy   finish
happen   like   listen   live   look   open   play   start
stay   stop   talk   test   try   travel   visit   walk   watch

2 Write these irregular verbs in the past simple tense.

be   become   do   go   get   have   know   leave
make   put   run   see   sit   sleep   take   tell

3 Complete this conversation with *did, didn't, was, wasn't, were* or *weren't*.

1 '___ you enjoy your holiday?' 'Yes, I ___ .'
2 '___ the weather good?' 'Yes, it ___ .'
3 '___ you go to the local museums?' 'Yes, we ___ .
   They ___ very interesting.'
4 '___ you send any postcards?' 'No, we ___ .'
5 '___ you buy anything?' 'No, we ___ have any money.'
6 '___ you pleased with the hotel?' 'Yes, we ___ .'

4 Write short answers to these questions about Thomas Cook. Give the correct information.

1 Did he live in Nottingham?
2 Was his first trip to Loughborough?
3 Did he take people to Manchester in 1845?
4 Did he open an office in London?
5 Were tourists interested in the Civil War battlefields?
6 Was it dangerous to carry large amounts of cash?

5 Write questions about Thomas Cook using the prompts below.

1 what/do?
2 what/be son's name?
3 where/go on first trip abroad?
4 when/publish the first travel magazine?
5 what/take to the Holy Land?
6 when/die?

6 Look back at the passage on page 16 and write answers to the questions in 5.
He was a printer.

7 Complete these sentences with *I, my, you, your, he, his, she, her, we, our, they* or *their*.

1 I know ___ face. Is she famous?
2 We know them quite well. ___ children go to the same school as ours.
3 Philip, this is ___ friend, Mary.
4 What do you do, Peter? What's ___ job?
5 I'll call you later. What's ___ telephone number?
6 That belongs to me. It's got ___ name on it.

SOUNDS

1 📼 Listen and say these words aloud.

<u>th</u>is <u>th</u>eatre ca<u>th</u>edral <u>th</u>ink <u>th</u>ank you <u>th</u>ey

Is the underlined sound /ð/ or /θ/? Put the words in two columns.

2 📼 Listen and say these words aloud.

n<u>o</u>t kn<u>o</u>w l<u>o</u>ts b<u>oa</u>t ph<u>o</u>t<u>o</u> vide<u>o</u> sh<u>o</u>p d<u>o</u>n't g<u>o</u>
wh<u>a</u>t disc<u>o</u> <u>o</u>pera <u>o</u>pening

Is the underlined sound /ɒ/ or /əʊ/? Put the words in two columns.

3 📼 Listen to the sentences. Put a tick (✓) if you think speaker *B* sounds friendly.

1 **A** Did you arrive late this morning? **B** No, I didn't.
2 **A** She didn't say hello. **B** Yes, she did.
3 **A** Did you give me your passport? **B** Yes, I did.
4 **A** Did they pay you for the tickets? **B** No, they didn't.

Now work in pairs and say the sentences aloud. Try to sound friendly.

WRITING

1 You are going to write a story called *The least successful annual conference*. Look at the questions below and try to guess the answers. Write full answers to the questions. Leave a blank when there is any information that you don't know.

> **THE LEAST SUCCESSFUL ANNUAL CONFERENCE**
> •
> **1 Where did the Association of British Travel Agents go for their annual conference in 1985?**
>
> **2 Why was the flight from Gatwick to Naples delayed?**
>
> **3 Why were many people ill?**
>
> **4 Who did the organisers ask to give a speech to the delegates in the forum at Pompeii?**
>
> **5 What did a local travel agent decide to do as a gesture of friendship?**
>
> **6 What happened as the Minister began his speech?**
>
> **7 Where did the flowers land?**
>
> **8 Why did no one hear the speech?**
>
> **9 How many times did the plane pass over Pompeii?**
>
> **10 Where did the roses land each time?**
>
> **11 What happened the last time it flew over the delegates?**

1 In 1985 the Association of British Travel Agents went to ___ for their annual conference.

2 Now turn to Communication activity 10 on page 99 and read the story. Fill in the blanks in your version.

SPEAKING

1 Find someone in your class who:
– left home early this morning
– travelled by plane last month
– had a holiday three months ago
– went on a bus yesterday
– walked to school/work ten days ago
– took a train last week
– stayed in a hotel last year
– went abroad in 1993

2 Draw your family tree, but include two 'false' relatives who don't exist. Think of some interesting or unusual information about the homes of each relative and invent information for the two relatives who don't exist.

3 Work in pairs. Ask and answer questions about your families. Try and guess who the 'false' relatives are.

11 *How ambitious are you?*

Verb patterns (2): *to* + infinitive; *going to* for intentions, *would like to* for ambitions

READING

1 How ambitious are you? Put a tick (✓) by the ambitions you have. Have you any other ambitions?

- stop work   – move house   – change your job
- learn to fly   – run a marathon   – learn a foreign language
- learn to ski   – work abroad   – be healthy and happy
- have children   – write a novel   – travel around the world
- go back to university   – earn a lot of money   – become famous
- live in the country   – go to Disneyland

2 Read *How ambitious are you?* and answer the questions.

3 Turn to Communication activity 9 on page 99 and find out how ambitious you are.

How ambitious are you?

1 Your neighbour buys a new Porsche. What do you say?
- *a* I don't need an expensive car.
- *b* One day I'm going to buy one, too.
- *c* I'd like to let the air out of its tyres.

2 Your boss leaves very suddenly. What do you think?
- *a* They're going to give me his/her job.
- *b* They're going to appoint someone else.
- *c* They're going to say it's my fault.

3 Ihich ambition do you have?
- *a* I'd like to be rich.
- *b* I'd like to be famous.
- *c* I'd like a drink.

4 You aren't happy with your job. What would you like?
- *a* More responsibility.
- *b* More money.
- *c* More weekend.

5 Your best friend is writing a novel. What do you say?
- *a* I'm going to write a best-selling novel.
- *b* I'm going to write a postcard.
- *c* What a coincidence! You're writing a novel and I'm reading one.

6 What does your future hold for you?
- *a* I'm going to be president.
- *b* I'm going to be happy.
- *c* I'm going to be late.

7 There's a marathon race on television. What do you say?
- *a* I'm going to do that next year.
- *b* I'd like to do that but I'm not very fit.
- *c* What's on the other channel?

8 Which of these statements do you agree with?
- *a* Every day, in every way, I'm getting better and better.
- *b* Tomorrow is the start of the rest of my life.
- *c* If you don't succeed, try again. Then give up.

GRAMMAR

> **Verb patterns (2):** *to + infinitive*
> **You can put *to* + infinitive after many verbs.**
> *I **want to** leave now.   He **decided to** drive to work.*
> *She's **learning to** fly.   She **needs to** pass her exam.*
>
> *Going to, would like to*
> **You can use *going to* + infinitive to talk about future intentions or plans which are fairly certain.**
> *I'm studying medicine. I'**m going to** be a doctor.*
> *I'**m not going to** be an accountant.*
>
> **You can use *would like to* + infinitive to talk about ambitions, hopes or preferences.**
> *I'**d like to** speak English fluently.   I **wouldn't like to** run a marathon.*
>
> **Remember that *like* + *-ing* means *enjoy*.**
> *I like learning English. = I enjoy learning English.*
>
> **For more information see Verb patterns (1) on page 9.**

1 Choose the correct verb pattern.

1 He's got a place at Essex University. *He would like to/He is going to* study there.
2 She's got her plane ticket and *she'd like to/she's going to* go to Canada.
3 *He'd like to/He is going to* buy a new car, but it's too expensive.
4 *I'd like to/I'm going to* work in television but there aren't many jobs.
5 She enjoys her job. *She wouldn't like to/She isn't going to* change it.
6 He's got a new job with a foreign company. *He'd like to/He is going to* work abroad.

2 Work in pairs and talk about the ambitions you ticked in *Reading* activity 1.
I'd like to run a marathon.

VOCABULARY AND WRITING

1 Match the verbs with the nouns in the box below.

abroad	change	earn	jobs	house	a foreign language	learn	a marathon
money	move	a novel	read	stop	run	study	work	write

Can you think of other nouns which can go with the verbs?

2 Look at this passage about Hanna's ambition. Find out:

– what she'd like to do
– why she'd like to do it
– what she needs to do to achieve it
– what she's going to do

> **H**anna is forty-five years old and is a technician in Ludwigshafen, Germany. She wants to learn to fly a light airplane because she loves flying. 'I'm always a passenger and I'd really like to know what it feels like to be in control,' she says. So she's taking flying lessons at her nearest private airport, and she's going to take her pilot's test in three weeks' time. 'It's not going to be easy, but I love it. It's going to change my life.'

3 Think about your ambitions. Make a list of what you'd like to do.
earn a lot of money, write a novel...

4 Choose one ambition, and make notes about why you'd like to do it.
earn a lot of money – buy a big house...

5 Now make notes about what you need to do to achieve it.
change my job, learn a foreign language...

6 Finally, make notes about what you're going to do to achieve your ambition.
look at job advertisements, go to evening class...

7 Work in pairs. Ask questions about your partner's ambition and take notes. Answer questions about your ambition.

8 From your notes, join what your partner wants to do and why with *because*.
Imogen would like to change her job because she wants to work abroad.

Join what your partner needs to do and what he/she's going to do with *so*.
She needs to decide where she wants to live, so she's going to travel round Europe.

9 Write a paragraph about your partner's ambition. Use the prompts and passage in 2 to help you.

12 English in the future

Will for predictions

VOCABULARY

1 Look at the words in the box and put them under two headings: *jobs* and *subjects*.

> accountant actor arithmetic banker biology chemistry computer science dancer doctor economics engineer geography history journalist languages maths nurse physics physical education politician secretary

2 Work in pairs and look at the lists you made in 1. Which jobs do you need English for? Which subjects do you need English in order to study? Are there any other jobs and subjects you need English for?

LISTENING

1 Think about learning English in the future in your country. Which of these predictions do you agree with? Put a tick (✓) if you agree and a cross (✗) if you disagree.

	You	Lynne	Greg	Your partner
Children will learn English from the age of six.				
There will be few adults who don't speak English.				
More lessons at school will be in English.				
Everyone will need to learn about British and American culture.				
Everyone will need English for their jobs.				
Everyone will learn English at home by television and computers.				
It will be more important to speak English than your own language.				

2 🔲 Listen to two English teachers talking about the statements. Put a tick (✓) if the speaker agrees with the statements, a cross (✗) if he or she disagrees and put ? if it's not clear.

3 Work in pairs. Can you remember what Lynne and Greg said?

🔲 Listen again and check.

GRAMMAR

> *Will*
>
> **You use *will* to make a prediction or express an opinion about the future.**
>
> *Children **will learn** English from the age of six.*
> *I think most people **will need** English for their jobs.*
> *I'm sure everyone **will speak** some English.*
>
> **You form the future simple with *will* + infinitive. You often use the contracted form *'ll*.**
>
> *At the end of the course I**'ll speak** English fluently.*
>
> **Negatives**
> *There **won't be** traditional language classes in school.*
> *There definitely **won't be** many teachers.*
>
> **Questions**   **Short answers**
> ***Will** we **use** English at work?*   *Yes, we **will**.*
>   *No, we **won't**.*

1 Think about the end of your course. Make predictions about your level of English. Use *I will* or *I won't*.

1 speak English fluently
2 be able to read an English newspaper
3 be able to understand radio broadcasts
4 be able to write reports in English
5 speak with a perfect English accent
6 be able to understand English songs
7 know a lot of vocabulary
8 use English for my work

2 Look at these predictions about jobs and studying in the future. Do you agree or disagree? Write questions to ask another student.

1 computers – replace secretaries, accountants
2 journalists – disappear – because no newspapers
3 economics – the most important school subject
4 teaching by television – very common
5 people – no need to study maths – because computers can do calculations
6 few people – need geography, history

1 Do you think computers will replace secretaries and accountants?

3 Now work in pairs and find out what your partner thinks.
Do you think computers will replace secretaries and accountants? No, I don't.

SOUNDS

1 Say these words aloud. Which words are stressed on the first syllable? Which words are stressed on the second syllable? Which words are stressed on the third syllable? Put them in three columns.

accountant actor arithmetic banker biology
chemistry computer economics education
engineer geography politician secretary

□◻	◻□◻	◻◻□◻
actor	*accountant*	*economics*

🔊 Listen and check.

2 🔊 Listen and say these words.

<u>a</u>ge l<u>e</u>sson ch<u>e</u>mistry <u>e</u>ducation s<u>e</u>cretary r<u>ai</u>lway
st<u>a</u>tion fri<u>e</u>nd holid<u>a</u>y

Is the underlined sound /e/ or /eɪ/? Put the words in two columns.

SPEAKING AND WRITING

1 In pairs, check your answers to *Listening* activity 2. Find out what your partner thinks and complete the *Your partner* column in the chart.

2 Find out what other people in your class think about the future of English and make notes.

3 Write a paragraph saying what people in your class think.
All of us think that children will learn English from the age of six. Most of us think that we'll use English for our jobs. Some of us think that learning English will be more important than our own language. Nobody thinks that all lessons at school will be in English.

Foreign travels

Going to for plans and *will* for decisions; expressions of future time

LISTENING

1 You're going to hear Duncan, an English student, talking to a friend, Cathy, about a visit he's going to make to South America. He's going to visit some of these places. Do you know which countries the places are in?

Places

The Amazon	☐	Lake Titicaca	☐	Rio	☐
Bel Horizonte	☐	Lima	☐	Santiago	☐
Buenos Aires	☐	Machu Picchu	☐	Sao Paulo	☐
Caracas	☐	Montevideo	☐	Valparaiso	☐
Cordoba	☐	Patagonia	☐		

2 Look at the map in Communication activity 21 on page 101 and check your answers. (No prizes if you live in South America!)

3 🔈 Listen to the conversation and number the places in 1 as Duncan mentions them.

4 Work in pairs. Try to remember what Duncan plans to do in each place. Put the number of the place by what he's going to do there.

lie on the beach	☐	spend a week in the jungle	☐
visit the ruins	☐	go round the museum	☐
do some sightseeing	☐	take the cable car up the mountain	☐
stay in a hotel	☐	buy some souvenirs	☐
meet his girlfriend	☐	hire a car	☐
travel by coach	☐	learn to dance the samba	☐

🔈 Now listen again and check.

5 🔈 Listen to the rest of the conversation. Underline the verbs used.

CATHY Well, have a nice time! Have you got a good guide book?

DUNCAN No, I haven't. But *I'll get/I'm going to get* one. It's on my shopping list of things to buy before I go.

CATHY Well, they say the best one is the South American handbook.

DUNCAN Really? Well, *I'll get/I'm going to get* it when I go into town.

CATHY Look, *I'll go/I'm going* into town right now because I need to do some shopping. *I'll buy/I'm going to buy* it for you at the bookshop, if you like.

DUNCAN Really?

CATHY Yes, of course.

DUNCAN Well, *I'll give/I'm going to give* you the money for it right now.

CATHY OK, and *I'll bring/I'm going to bring* it round tonight. Who knows, perhaps *I'll borrow/I'm going to borrow* it from you one day.

DUNCAN OK. Thanks very much.

GRAMMAR

> ### Going to for plans
> You use *going to* to talk about things which are arranged or sure to happen.
> *I'm going to visit South America.   I'm going to visit Buenos Aires.*
>
> You use *going to* for decisions you made before the moment of speaking.
> *I'm going to buy a guide book.*
>
> ### Will for decisions
> You use *will* for decisions you make at the moment of speaking.
> *I'll give you the money right now.*
>
> You often use *will* for offers. (For more information about offering, see Lesson 30.)
>
> You usually use the present continuous with *go* and *come.*
>
> | | | |
> |---|---|---|
> | *He's going to South America.* | not | *He's going to go to South America.* |
> | *She's coming with us.* | not | *She's going to come with us.* |
>
> ### Expressions of future time
> You can use these expressions of future time to say when you are going to do things.
>
> | | |
> |---|---|
> | *tomorrow* | *morning   afternoon   evening* |
> | *next* | *week   month   year* |
> | *in* | *two days' time   three months' time   five years' time* |

1 Does Duncan make his decisions before or at the moment of speaking to Cathy? Which verb form do you use to talk about decisions? Which one do you use to talk about plans?

2 Work in pairs and check your answers to *Listening* activity 4. Which things is Duncan sure he is going to do, and which things is he not sure will happen?
He's going to fly to Rio. He'll probably take the cable car up the mountain.

3 Choose the correct verb form.
  1 'I need a strong bag.' '*I'm going to/I will* get you one.'
  2 I bought a good map, because *I'm going to/I will* go to South America.
  3 'Where *will you /are you going to* stay?' '*We'll /We're going to* stay with friends, probably.'
  4 I need some fresh air. Perhaps *I'm going to/I'll* have a walk in the park.

4 Choose six expressions of future time and write three sentences about things you're sure you're going to do, and three things you're not sure will happen.
I'm going to have a holiday next month.
I'll probably move abroad in five years' time.

SPEAKING AND VOCABULARY

1 Work in groups of two or three. You're planning a trip somewhere. Decide where you'd like to go.

2 In your groups, look at the words in the box and decide which things you will need for the trip. Add three or four more things.

> aspirin backpack map camera
> currency food guide book
> handbag medical kit passport
> penknife razor scissors tent
> sleeping bag suitcase wallet
> toothbrush toothpaste watch
> traveller's cheques walkman

3 Tell the others in your group which things you'll get.
I'll get the map.
OK, I'll buy some food.
Yes, and I'll bring my walkman.

4 Check who is going get which things.
So, Mario is going to get the map and Tanya is going to buy some food.
Thanks for your help Paco, but...

14 *In Dublin's fair city*

Prepositions of place; asking for and giving directions

VOCABULARY AND READING

1 Work in pairs. Look at the map of Dublin, the capital city of Ireland. Can you see any of these town features on the map? What are the places called?

> bank   bus station   college
> hospital   law courts   library
> parliament   pub   railway station
> river   square   town hall

The Bank of Ireland...

2 The article you are going to read is about the pubs in Dublin. Which of these words can you use to talk about pubs or bars in your country?

> traditional   conversation   jokes
> old-fashioned   welcome   private
> charm   modern   beautiful   local
> famous   thirsty   lively   literary
> sophisticated   elegant   left-wing

3 Read *In Dublin's fair city* and decide which pub(s) you'd like to visit and why.

LISTENING

1 🔲 Listen to a tour guide at the start of a tour. Where is he standing?

2 🔲 Listen to the guide explaining the route of the tour. Draw the route on the map.

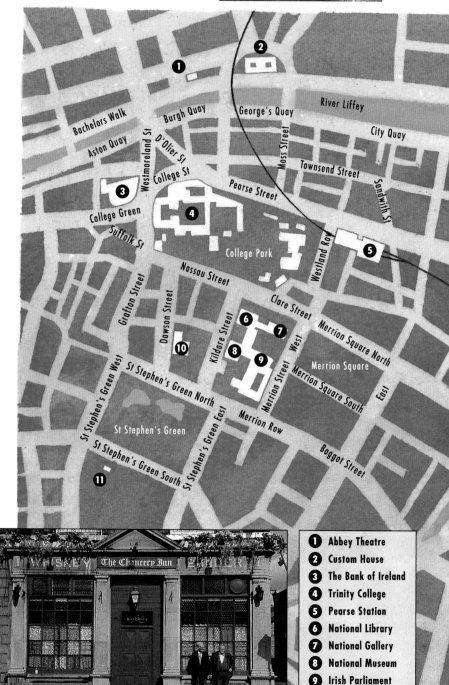

1. Abbey Theatre
2. Custom House
3. The Bank of Ireland
4. Trinity College
5. Pearse Station
6. National Library
7. National Gallery
8. National Museum
9. Irish Parliament
10. Mansion House
11. University Church

IN DUBLIN'S FAIR CITY

'I went into a pub one morning and asked for a Guinness. The barman said, "Sorry, we're closed. We open in fifteen minutes." Then he paused. "But would you like a drink while you're waiting?"'
– *An English visitor to Dublin*

Dublin means many things to many people. To some it is a city of writers, the city of Jonathan Swift, Oliver Goldsmith, James Joyce and W B Yeats. For others, it is the city of talkers, its pubs full of Guinness and jokes, and the source of the writers' inspiration.

You can still find many traditional pubs in Dublin and they're open all day from 10.30am on Mondays to Saturdays and most of the day on Sunday. You're sure of a warm welcome, especially if you offer to buy the locals a drink. Here are a few to visit in the centre.

Doheny and Nesbitt's in Baggott Street is a lively, old-fashioned pub, long and narrow, with a small private bar at each end. Government ministers, civil servants and journalists use it for meetings.

The Horsehoe Bar of the elegant Shelbourne Hotel on St Stephen's Green keeps its charm and intimacy in its modern, sophisticated surroundings.

Mulligan's in Poolbeg Street opened in 1872 and has a number of sections: students in the entrance on the left, journalists and TV people on the right, and left-wing politicians in the far bar.

Neary's, in Chatham Street, is a beautiful, old-style pub visited by actors from the Gaiety Theatre.

The Bailey, in Duke Street, is a very traditional pub, and famous for its literary connections. It was in James Joyce's Ulysses, where it is known as Burton's.

There are hundreds of other pubs as well, where visitors can enjoy something of the real Dublin. You don't need to be thirsty to enjoy the pub; just enjoy the conversation!

GRAMMAR AND FUNCTIONS

> **Prepositions of place**
> *The Bank of Ireland is **opposite** Trinity College.*
> *Pearse Station is **next to** Trinity College.*
> *College Park is **in** Trinity College.*
> *Dawson Street is **between** St Stephen's Green and Nassau Street.*
> *The National Library is **behind** the National Museum.*
> *The National Museum is **on the corner of** College Green and Westmoreland Street.*
> *The National Museum is **in front of** the National Library.*
>
> **Asking for and giving directions**
> *How do I get to ...?*   *Go straight ahead/on.*
> *Go down/along...*   *Go to the end of...*
> *Cross over...*   *Turn left/right into...*
> *It's on the left/right.*   *It's at the crossroads.*
> *Take the first/second turning on the left/right.*

1 Look at the map and say where these places are. Use the prepositions in the grammar box.

The Irish Parliament   the Mansion House
Merrion Square   Kildare Street   The National Gallery

2 Write directions for the tour you heard in *Listening* activity 2. Use the prepositions in the grammar box, and the map to help you.

🔘🔘   Listen again and check.

SOUNDS

1 Say these words aloud.

bank pub map article Dublin number others
charm far bar actor bus national

Is the underlined sound /ɑː/ , /æ/ or /ʌ/? Put the words in three columns.

🔘🔘   Listen and check.

2 Underline the stressed words in these questions.

1 Excuse me, how do I get to the post office?
2 Excuse me, is there a bank near here?
3 Excuse me, could you tell me the way to the station, please?
4 Excuse me, where's the nearest pub?

🔘🔘   Listen and check. Ask the questions aloud.

SPEAKING

1 Work in pairs.

Student A: Think of a well-known place in the town where you are now. Imagine you are looking at it and describe where you are to Student B. Don't say what the place is.

Student B: Listen to Student A describing a well-known place in the town where you are now. Guess what the place is.

Change round when you're ready.

2 Work in pairs.

Student A: Turn to Communication activity 7 on page 99.
Student B: Turn to Communication activity 15 on page 100.

15 *An apple a day*

Expressions of quantity (1): countable and uncountable nouns, *some* and *any*, *much* and *many*

VOCABULARY

1 What sort of things do you eat, drink or use in your cooking? Look at the words in the box and put them in lists under these headings: *every day, twice a week, every week, on special occasions, never*.

> apples bananas beef beer biscuits bread butter cabbage carrots cheese
> chicken coffee eggs fish fruit grapes ham juice lamb lettuce meat
> milk oil onions oranges pasta peaches peas pork potatoes rice salad
> strawberries tea tomatoes vegetables water wine

Can you think of two or three more things to add to each list?

2 Work in pairs and compare your lists. Which of the things can you see in the photo?

3 Look at these words.

| bottle cup glass kilo· loaf packet piece slice tin |

a **bottle** of beer   a **cup** of tea   a **glass** of water
a **kilo** of potatoes   a **loaf** of bread
a **packet** of biscuits   a **piece** of cheese
a **slice** of bread   a **tin** of peaches

Which other items in the vocabulary box can you use with these words?

GRAMMAR

> Expressions of quantity (1): countable and uncountable nouns
>
> **Countable nouns have both a singular and a plural form.**
> *an apple – two apples   a peach – two peaches*
> **Uncountable nouns do not usually have a plural form.**
> *bread, beef, butter, coffee, water*
>
> *Some* and *any*
> **You usually use *some* in affirmative sentences.**
> *I'd like an orange, two apples, **some** peaches and **some** water.*
> **You usually use *any* in negative sentences and questions.**
> *We haven't got **any** butter.   Are there **any** eggs?*
>
> *Much* and *many*
> **You usually use *much* and *many* in negative sentences and questions. You use *many* with countable nouns.**
> *We haven't got **many** carrots.*
> *How **many** eggs would you like?*
> **You use *much* with uncountable nouns.**
> *There isn't **much** cheese.   How **much** butter do you need?*

1 Look at the words in the vocabulary box again. Write
C (countable) or U (uncountable) by them.

2 Complete the dialogue with *some, any, much* or *many.*

  A We need ___ water. How ___ bottles do we need?
  B Two. And we haven't got ___ fruit. Shall we get ___ peaches?
  A OK. Have we got ___ tea?
  B No, how ___ do we need?
  A Just a packet.

  [cassette]   Now listen and check.

LISTENING AND SPEAKING

1 Work in groups of three. You are going to hear Karen, who lives in Hong Kong, and Pat, who lives in San Francisco, talking about a typical breakfast, lunch and dinner. First, make sure you understand these food items they mention:

cereal   pie   steamed dumplings   toast
sandwich   chowder

2 *Student A:*   Turn to Communication activity 1
                    on page 98.

  Student B: Turn to Communication activity 12
                    on page 100.

  Student C: Turn to Communication activity 20
                    on page 101.

3 Now work together and complete the columns *Karen* and *Pat.*

	Karen	**Pat**	**You**
Typical breakfast			
Typical lunch			
Typical dinner			

4 What do you have for a typical breakfast, lunch and dinner? Complete the *You* column in the chart.

5 Compare the speakers' typical meals with your typical meals. Use these expressions.

a lot of/lots of   quite a lot of   a few /a little
not much/many   hardly any   not any

Karen eats a lot of meat, and so do I. Pat eats hardly any vegetables, but I eat lots.

6 Find out what sort of things other people in your class eat, drink or use in their cooking.
Fadia, do you drink tea?   *Yes, I do.*

Then find out how much they eat, drink or use.
How much tea do you drink every day? How many cups a day do you drink?

Progress check **11-15**

VOCABULARY

1 Word maps are a good way of remembering and organising new vocabulary. Make a word map of your town or city.

good food — my town — pubs

interesting people

cheap housing

nice buildings

2 Look at these words for jobs. Which come from verbs? Write the verb.

actor banker dancer footballer journalist
manager musician politician teacher writer

actor – to act

Which ones come from other nouns? Write the nouns.

banker – a bank

You can make other words using a suffix. Look at these suffixes.

*act**or** bank**er***

Underline all the suffixes in the words above. Words with these suffixes are often jobs.

3 Look at the vocabulary boxes in lessons 11 – 15 again. Choose words which are useful to you and group them under headings of your choice in your *Wordbank*.

GRAMMAR

1 Write questions about Jane with *going to*.

1 be an accountant
2 live in the USA
3 start her own company
4 learn Italian
5 visit South America
6 start a new life

1 Is she going to be an accountant?

2 Write answers to the questions you wrote in 1.

1 be a doctor
2 move to Spain
3 work in a hospital
4 learn Spanish
5 travel round Spain and Portugal
6 stay in contact with her old friends

1 Is Jane going to be an accountant?
No, she isn't. She's going to be a doctor.

3 You're going to start a new, exciting job tomorrow. Say what you are or aren't going to do.

– walk to work   – arrive early   – drink less coffee
– work hard   – be friendly   – stay late

4 Complete these sentences with *(be) going to* or *would like to*.

1 He's got his ticket and he _____ fly to Madrid.
2 I _____ buy a car, but I haven't got any money.
3 It _____ rain today.
4 We _____ go on holiday but we're too busy.
5 He's got his coat on and he _____ leave now.
6 We sold our flat last week and we _____ live abroad.

5 Complete the sentences with *will* or *(be) going to*.

1 It's very early. Maybe I ___ go back to bed.
2 He ___ fly to Bel Horizonte next week.
3 She ___ have a baby next July.
4 'I'm so tired.' 'I ___ take you home by car.'
5 He works in Lisbon, so he ___ move there.
6 Maybe we ___ have dinner in a restaurant.

6 Make predictions about the following.
Use *I think/perhaps/maybe + will*.

1 next weekend
2 your job
3 next holidays
4 your family
5 next year
6 your friends

7 Write sentences saying where these places are in your town. Use these prepositions:

at on opposite next to in between behind
on the corner of   in front of

1 the post office
2 the bank
3 the library
4 the supermarket
5 the swimming pool
6 the cinema
7 the football stadium
8 the park

8 Write directions from where you are now to the following places.

1 the railway station
2 the bus station
3 a petrol station
4 the cinema

9 Are these things countable or uncountable? Write *C* or *U*.

egg money orange juice apple sugar potato butter rice strawberry cheese

10 Write sentences using *How much ... have you got?* or *How many ... have you got?* and the words in 9.
How many eggs have you got?
How much money have you got?

11 Complete these sentences with *some* or *any.*

1 Have you got ___ oranges?
2 I'd like ___ wine please.
3 I don't have ___ money with me.
4 Is there ___ water?
5 We've got ___ chicken but we haven't got ___ salad.
6 I'll get you ___ bread, if you like.

SOUNDS

1 Say these words aloud. Underline the /ə/ sound.

pizza polite police theatre cinema opera performance dinner weather

[cassette icon] Now listen and check.

2 Say these words aloud.

charm chicken politician traditional cheese she old-fashioned peach fish

Is the underlined sound /tʃ/ or /ʃ/? Put the words in two columns.

[cassette icon] Listen and check.

3 Look at this true sentence.
Joe is going to study maths at university.

[cassette icon] Listen and answer the questions below with the true sentence. Change the stressed word each time.

1 Is Tim going to study maths at university?
2 Is Joe going to study physics at university?
3 Is Joe going to study maths at school?
4 Did Joe study maths at university?

*1 No, **Joe** is going to study maths at university.*

4 [cassette icon] Listen to these questions. Put a tick (✓) if the speaker sounds polite.

1 How do I get to the station?
2 How do I get to the hospital?
3 Could you tell me where the town hall is?
4 Could you tell me where the bus station is?
5 Where's the post office?
6 Where's the river?

Now say the questions aloud. Try to sound polite.

SPEAKING

1 Work in pairs. One of the other pairs is coming for lunch. Decide:
– what dish you'll make
– where you'll have lunch.

2 Write a note to another pair. Say:
– where you're going to have lunch
– how to get there.

3 You need to buy the ingredients for the dish you chose in 1. Write each ingredient on a piece of paper. With your partner, discuss which ingredients you'll get.

A *I'll get the tomatoes.*
B *And I'll buy some onions.*
A *OK. You're going to buy some onions and I'm going to buy the tomatoes.*

4 Give the pieces of paper with ingredients to your teacher, and you will receive some different ingredients. Go round asking other people if they have got the ingredients that you're going to get, and saying what you've got.

Have you got any tomatoes?
Yes, I have./No, I haven't.

Give your ingredients to anyone who asks for them, even if you need them! The first pair to get all their ingredients is the winner.

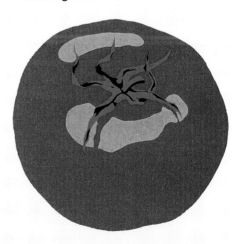

16 | *What's on?*

Prepositions of time and place; making invitations and suggestions

VOCABULARY

Look at the words in the box. Which words go under these headings: *What's on? Where?*

> ballet cinema closing time club concert disco
> exhibition film gallery interval match musical
> museum opening hours opera opera house painting
> performance play row sculpture seat stadium
> theatre ticket

What's on?	*Where?*
ballet	*theatre*
film	*cinema*

Now put the remaining words in a third column: *Related words*.

What's on?	*Where?*	*Related words*
ballet	*theatre*	*ticket, row, seat*

LISTENING AND SPEAKING

1 Work in groups of three. You are going to hear people talking about typical entertainment in Japan and Argentina. Ken talks about *karaoke* in Japan, and Philippa talks about *tango* in Argentina.

Student A: Turn to Communication activity 2 on page 98.
Student B: Turn to Communication activity 19 on page 101.
Student C: Turn to Communication activity 11 on page 100.

2 How much do you know about karaoke and tango now? Work together and complete the chart.

	karaoke	**tango**
Type of entertainment		
Place of entertainment		
Performers		
Type of music		
Reasons why people enjoy it		

3 Talk about a typical type of entertainment in your country. Use the chart to help you.

GRAMMAR AND FUNCTIONS

> **Prepositions of time and place**
> **at**   *the Dominion Theatre   8pm   the football match*
> **on**   *Monday   31st July*
> **in**   *June   1996   London   England*
> **to**   *go to work   go to the cinema   go to a party*
>
> **Making invitations and suggestions**
> **Would you like to** *come to the cinema?*
> **How about** com**ing** *to the cinema?*
> **Let's** *go to the cinema.*
>
> **Accepting**   **Refusing**
> *I'd love to.*   *I'm sorry, I can't. I'm busy.*

1 Look at the prepositions of time and place in the grammar box. Write the phrases in two columns: *time* and *place*.

2 Complete the rules about when we use *at*, *on* or *in* to talk about time.

1 You use ___ to talk about months and years.
2 You use ___ to talk about days of the week and dates.
3 You use ___ to talk about a point of time in the day (eg *eight o'clock*).

3 Complete these sentences with *at, in, on* or *to*. Sometimes you can use more than one preposition. Is there a difference in meaning?

1 The match starts ___ 3pm ___ Saturday.
2 ___ the National Gallery ___ June and July there's a Van Gogh exhibition.
3 I'd like to go ___ the opera when we're ___ Moscow.
4 ___ Paris there's a World Cup football match ___ Saturday.
5 I'm going ___ a party ___ 20th June.
6 Would you like to come ___ the theatre ___ London ___ Sunday?

4 Make a list of what's on in your town at the moment. Go round the class inviting people to go out with you on different days. Accept or refuse their invitations, and try to fill up your diary.
Would you like to come to the match on Friday?
Yes, I'd love to!
How about coming to the cinema on Saturday?
I'm sorry, I'm busy. How about Sunday?

WRITING

1 Look at the formal invitation below.

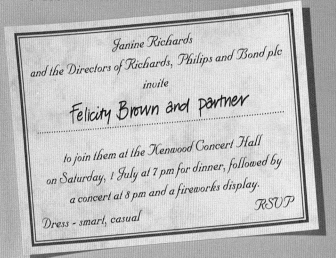

> *Janine Richards*
> *and the Directors of Richards, Philips and Bond plc*
> *invite*
> **Felicity Brown and partner**
> *to join them at the Kenwood Concert Hall*
> *on Saturday, 1 July at 7 pm for dinner, followed by*
> *a concert at 8 pm and a fireworks display.*
>   *RSVP*
> *Dress - smart, casual*

Work in pairs. Invite your partner to go with you. Complete the letter.

> *45 Thame Street*
> *Garsington, OX7 1LS*
> *3rd June*
>
> *Dear* _____
>
> *I've got an invitation from*
> _____
> *to join them on* _____
> *for* _____ *followed by a* _____
> *and a* _____ *. Would*
> *you like to come? It starts at* _____ *.*
>
> *Best wishes*
>
> _____

2 Give your partner your invitation. Now reply to his/her invitation. Use these phrases to help you.
Thank you for the invitation to ... I'd love to come.
Shall we meet at...   I'm afraid I'm busy.

17 | *Famous faces*

Describing people

VOCABULARY

1 Work in pairs and look at the words in the box. Which are adjectives and which are nouns?

> attractive bald beard beautiful black blonde brown curly
> dark face fair fat glasses good-looking hair head kind
> long man middle-aged moustache nice old pretty
> medium-height round short shy slim square straight tall
> teenager thin ugly woman young

Group any nouns and adjectives which often go together.
attractive face...

Which adjectives can you use to talk about the following?
– height   – age   – looks   – build   – character

2 Think of a famous person. Choose three or four words from the vocabulary box which you can use to describe his/her appearance.

Now tell your partner the name of your famous person. He/she must guess which words you chose.
My famous person is Mickey Mouse.
Did you choose short, dark, middle-aged?
Yes... and bald!

FUNCTIONS

> **Describing people**
> **Appearance**
> **You use *look like* to describe people's appearance.**
> ***What** does she **look like**? She's tall and she's got fair hair.*
> *She **looks like** a banker.*
> ***Who** does he **look like**? He **looks like** his father.*
>
> **Character**
> **You use *be like* to describe people's character.**
> ***What's** he **like**? He's nice.*
> ***Who's** he **like**? He's **like** his father.*
>
> *She's **quite** nice.   He's **about** twenty.   She's **about** one metre sixty.*
> *He's **very** tall.   He's **in** his **mid**-twenties.   She's **in** her **mid**-thirties.*
> *He's **really** handsome.   She's about thirty, **with** dark hair.*

WRITING

1 You are going to the station to meet someone who doesn't know you, and you're going to write a letter describing your appearance. First of all, make notes about these aspects of your appearance:
– age – height – looks – hair – build

age: mid-twenties   *height: quite tall*

2 Now write sentences describing your appearance.
I'm in my mid-twenties. I'm quite tall.

3 Then join the sentences using *and*.
I'm in my mid-twenties and I'm quite tall.

4 Write a letter describing what you look like.

> (Write your address here)
>
> (Write the date here)
>
> Dear Mr. Freeman,
>
> I am looking forward to meeting you at the station next Monday. I will be there at ten o'clock and will wait for you on the platform.
>
> (Describe your appearance)
>
> Yours sincerely,
>
> (Write your full name here)

1 Complete the sentences with *like* if necessary.

1 What does she look ___ ? She looks ___ very kind.
2 Who's she ___ ? She's ___ her mother.
3 What's he ___ ? He's ___ lovely!
4 Who does he look ___ ? He looks ___ his brother.

2 Match the questions and the answers.

1 How old is he?   a One metre seventy-eight.
2 How tall is he?   b Blond.
3 What colour is his hair? c He's quite young, good-looking and slim.
4 What does he look like? d Twenty-one.

3 Look at the sentences in the grammar box above. Write four sentences describing people you know with *quite, very* and *really.*
My father is very old and really intelligent.

SPEAKING

1 Work in pairs. Choose someone in the pictures and describe him or her to your partner. Can they guess who you're describing?

2 Talk about what you imagine the people in the pictures are like. You can use the words in *Vocabulary* activity 1 to describe them. Which person do you think you'd like to meet? Explain why.

5 Write five true pieces of information about your appearance and two false pieces. Show them to another student. Can he/she find the false information?

18 *Average age*

Making comparisons (1): comparative and superlative adjectives

VOCABULARY

1 Work in pairs. Choose five words to describe yourself. Use a dictionary if necessary.

> careful interesting clever cold
> confident fit funny imaginative
> intelligent kind lazy nervous
> optimistic patient pessimistic
> polite quiet calm rude sad
> sensitive nice serious tidy
> thoughtful

Think of other words you can use.
honest, friendly...

Discuss your choice of words with your partner.
I think I'm usually optimistic. And I'm always polite!

Does he/she agree with you?

2 Think of three people you admire very much. They can be politicians, musicians, sports personalities etc. or people you know personally. Choose the person you admire most and think of three adjectives to describe this person.

Then choose the second and third person you admire and think of three more adjectives for each person to explain why.

Now turn to Communication activity 13 on page 100.

READING

1 Read *Average age* and find things which are different from your experience or from the experience of people you know.

A v e r a g e   a g e

10 Ten is the year of the closest friendships – though not with the opposite sex. It is also the year when relationships with particular people or groups is strongest. The ten-year-old usually gets on well with parents but needs more time alone. Personal talents begin to show.

In the United States twenty is the average age for the first marriage for women, although probably only a third marry at this age because they want to; the others marry because of social pressure. The human brain is at its finest at twenty. It is the age when people can vote in Denmark, Japan, Norway and Switzerland. And in Japan it is the minimum age for buying alcohol. **20**

30 For optimists thirty is one of the happiest ages, for pessimists it marks the end of feeling young. At this age you need to take a little more care with your body than when you were younger. Young people who enjoyed an all-night party at twenty will feel much worse the next day at thirty. In Britain it is the youngest you can become a bishop.

Forty is the year of the 'middle-aged', although nobody who is forty wants to admit the fact. Bob Hope said that you are middle-aged when your age starts to show around your middle. In fact, the body starts to get smaller at forty and continues to do so until you die. **40**

50 Fifty is an age when old friendships get closer and relationships with colleagues and relatives warmer. According to old proverbs, fifty is the age when you should be rich. George Orwell said, 'At fifty everyone has the face he deserves.' People need to wear glasses and some food loses its strong taste.

Adapted from The Book of Ages, by Desmond Morris

2 Work in pairs. Which is the most surprising piece of information in the passage?

GRAMMAR AND FUNCTIONS

> ### Making comparisons (1)
> #### Comparative adjectives
> **You form the comparative of most adjectives by adding -er, -r, -ier or more + adjective.**
> *kind**er** nice**r** laz**ier** sad**der** **more** careful*
> *Fifty is an age when old friendships get **closer**.*
>
> #### Superlative adjectives
> **You form the superlative of adjectives with -est, -st, -iest or most + adjective.**
> *kind**est** nice**st** laz**iest** sad**dest** **most** careful*
> *For optimists thirty is one of **the happiest** ages.*
> **There are some irregular comparative and superlative forms.**
> *good better best   bad worse worst*

1 Look at the comparative and superlative adjectives in the box above. Write down the adjective they come from and the comparative and superlative forms.
kind   kinder   kindest

What's the rule for forming the comparative and superlative forms of short adjectives ending in -e, -y, a vowel + consonant?

What's the rule for forming the comparative and superlative forms of longer adjectives?

2 Make comparative and superlative adjectives from the following.

cold   imaginative   intelligent   fit   tidy   beautiful
polite   patient   young   funny   nervous   warm   old

3 Work in pairs and compare the information in *Average age* with your own experience.
In my country, I think that many women are older when they get married.

4 Work in groups of two or three. Talk about the best age for doing the following things.

– getting married   – buying a home
– having children   – leaving your parents' home
– going to university   – learning to drive
– leaving school   – learning a foreign language

I think it's best to get married at twenty-five.

Now write sentences comparing your opinions.
We all think it's best to leave school at sixteen.
Jerome thinks it's best to learn a foreign language at six, but we think it's best when you're older.

SPEAKING AND WRITING

1 Choose a superlative adjective you made in *Grammar* activity 2 to complete these questions about exceptional people, places and things.

– Who is the _____ person you know?
– What is the _____ thing you own?
– What is the _____ time of the year?
– Where is the _____ place you know?
– What is the _____ country for a holiday?
– Who is the _____ person to be with at a party?

Who is the funniest person you know?

2 Work in groups of three or four. Ask and answer the questions you wrote in 1 about exceptional people, places and things.
I think the funniest person I know is Dietrich.

3 Write sentences comparing your partners' answers to your questions.
Hans thinks the funniest person he knows is Dietrich.

19 | *Dressing up*

Making comparisons (2): *more than, less than, as...as*

VOCABULARY

1 Look at these words for clothes. Which do you wear? Put the words in four lists under these headings: *always, often, sometimes, never.*

| blouse coat dress hat jacket jeans shirt shoes |
| skirt socks suit sweater swimsuit tie tights |
| trainers trousers T-shirt underwear |

always: trousers   often: jeans   sometimes...

Think of other clothes you wear in winter, in summer, for work and at home and add them to your lists.

2 Choose suitable adjectives from the list below to describe your own clothes. Add them to the four lists you made in 1.

| black blue brown casual formal green grey |
| orange pink red smart white yellow |

always: dark trousers   often: smart jeans sometimes...

3 Work in pairs. Ask and say what you always, often, sometimes and never wear.
Do you often wear trousers, Erina?
Yes. I usually wear dark trousers.

READING

1 Read *Dressing up*, which is about clothing in Kuwait, Sweden and India, and find out if it says anything about:

– clothes for work
– traditional dress
– young people's fashions

Check your answers with another student.

In Kuwait, men and women wear their traditional dress most of the time. For men, it is a long robe and a cloth covering the head. For women it's similar and they wear a veil. Foreign male visitors usually wear lightweight cotton trousers and white shirts with short or long sleeves. Men often wear sandals during the day but never in the office. They wear a jacket and a tie for social occasions, but when it's really hot, it's usual to take off the jacket. Foreign women visitors usually wear long, loose clothes which cover their neck and arms.

The Swedish are very interested in clothes and are less formal than they were. People usually dress well in public and wear bright colours. In Sweden the winters are very cold, so overcoats and ski jackets are very common. Men wear business suits for work, with a shirt and a tie and women often wear trousers. People often carry a spare pair of shoes because you need boots outside. Children and teenagers are more casual than their parents. For school, they wear blue jeans and T-shirts.

Traditional dress in India for women is the *sari* and for men the *achkan* suit. The sari has its own distinctive style depending on which part of India it comes from – every region has its own special colours, decoration and style. The men wear their heavy and expensive *achkan* suits on formal occasions but for less formal occasions they wear the *kurtha* suit, a long shirt and loose trousers, which is not as heavy as the *achkan*. Indian people wear lighter colours as they grow older, and at funerals, white is the usual colour to wear.

Many people wear western-style clothes. For work they wear smart clothes, but not suits and ties. Women usually wear trousers and blouses but not dresses. Young people are as casual as young people all over the world with their jeans and T-shirts.

2 Work in pairs. Ask and say what clothes people in your country wear. Talk about clothes for work, clothes at home, traditional dress, young people's fashions.
Men wear suits for work. What do you wear?

3 Are the conventions for clothing in the passage different from conventions in your country?
In my country men don't usually wear sandals.

GRAMMAR AND FUNCTIONS

> Making comparisons (2): *more than, less than, as...as*
> *Children wear **more** casual clothes **than** their parents.*
> *They're **less** formal **than** they were.*
> *My father wears cheaper clothes **than** my mother.*
> *They're **as** casual **as** teenagers are all over the world.*
> *Dresses are **not as** popular **as** in Western countries.*
> *It's **the same as** my country.*
> *It's **different from** my country.*

1 Complete these sentences with *as, than* or *from*.

1 He's much smarter ___ I am.
2 She's ___ intelligent ___ he is.
3 Her clothes are different ___ mine.
4 She's got the same shoes ___ I have.
5 Sandals are less common here ___ in Kuwait.
6 Children are more casual ___ their parents.

2 Agree with these statements using *more* or *less* and the adjective in brackets.

1 He's less casual than she is. (formal)
2 It's more noisy now than it was. (quiet)
3 Clothes are cheaper here than at home. (expensive)
4 He's less optimistic than she is. (pessimistic)
5 It's easier to get good clothes here. (difficult)
6 He's less confident than she is. (nervous)

1 Yes, he's more formal than she is.

3 Write four sentences comparing what you wear and your appearance with other people. Use comparative adjectives.

I wear more colourful clothes than my father.

SOUNDS

1 Say these words aloud. Is the underlined sound /ə/ or /ɪ/?

small<u>e</u>r small<u>e</u>st bigg<u>e</u>r bigg<u>e</u>st clos<u>e</u>r clos<u>e</u>st happi<u>e</u>r happi<u>e</u>st funni<u>e</u>r funni<u>e</u>st

🔊 Listen and check.

2 🔊 Listen to the sentences in *Grammar and functions* activity 1. Notice that *than* is pronounced /ðən/, *as* is pronounced /əz/ and *from* is pronounced /frəm/.

Now say the sentences aloud.

3 🔊 Listen to someone disagreeing with the statements in *Grammar* activity 2.

Now work in pairs and disagree with the statements. Stress *more* or *less*.

LISTENING AND SPEAKING

1 Read these statements. Decide if they are true or false for your country.

	My Country	Britain
1 The weather is usually rather cold.		
2 It's difficult to buy good clothes.		
3 Good clothes are very expensive.		
4 People are quite formal.		
5 Many people are quite small.		
6 The quality of clothes design is good.		

2 🔊 Listen to Graham, from Britain, talking about clothing. Does he think the statements in 1 are true or false?

3 Work in pairs and answer the questions.

In your country...

– is the weather hotter or colder than in Britain?
– is it easier or more difficult to buy good clothes?
– are clothes cheaper or more expensive?
– are people more formal or more casual?
– are people smaller or larger than the British?
– is the quality of clothes design better or worse?

We have cold weather too. In my country it's more difficult to buy good clothes.

20 *Memorable journeys*

Talking about journey time, distance, speed and prices

VOCABULARY AND LISTENING

1 How do you say these numbers?

505
  a five hundred five
  b five hundred and five

478
  a four hundred and seventy eight
  b four hundred seventy eight

3,563
  a three thousand and five hundred sixty three
  b three thousand, five hundred and sixty three

45,781
  a forty five thousand, seven hundred and eighty one
  b forty five thousand, seven hundred eighty and one

▣ Now listen and check.

2 Say these numbers aloud.

346 678 2,345 18,664
24,589 123,456 202 54,566
3,481 407 10,020

▣ Now listen and check.

3 Work in pairs and look at the photo. Use these words to describe what you can see.

> arrive border cost desert distance
> drive driver gallon gas station get
> highway hill leave mile motel
> mountain move home passenger
> petrol police patrol reach set off
> speed limit take ticket truck turn off

Would you like to be one of the passengers?

4 ▣ Listen to Sarah, an English woman, talking about a memorable journey she made in the USA. As you listen, look at the vocabulary box again and tick (✓) the words you hear.

5 Tick (✓) the information you heard.

Journey time	9 days	19 days	90 days
Distance	250 miles	2,050 miles	2,500 miles
Price of petrol	25 cents a litre	50 cents a litre	55 cents a litre
Speed limit	50 miles an hour	55 miles an hour	70 miles an hour
Price of hotel rooms	$20 to $35 per person	$25 to $40 per person	$30 to $50 per person

▣ Now listen again and check.

46

SOUNDS

1 Say these words aloud. Underline the stressed syllable.

thirty thirteen fourteen forty seventeen seventy
nineteen ninety thirteen dollars fourteen kilometres
seventeen hours nineteen miles

▱▱ Listen and check. Which words are stressed on the first syllable? Which words are stressed on the second syllable?

2 ▱▱ Listen and write down the numbers you hear.

Now say the numbers aloud.

FUNCTIONS

Talking about journey time, distance, speed and prices

Journey time

How long does it take **by** car? (It takes) nine days.
How long does it take **by** train? Five hours.
How long does it take **on** foot? It's a five-minute
                                           walk/drive/flight/journey.

Distance

How far is it? It's 2,500 miles **away**.
How far is your school from your home? Ten kilometres.

Speed

How fast can you drive? Fifty-five miles **per** hour. (mph)
                              Ninety kilometres **an** hour. (km/h)

Prices

How much is petrol? (It's) $2 **a** gallon.
How much does petrol cost? (It costs) 25 cents **a** litre.
How much are hotel rooms? $25 **per** person **per** night.
How much do hotel rooms cost?

---

1 Work in pairs and check your answers to *Vocabulary and listening* activity 5.

2 Imagine these are the correct answers to questions about your country. Write the questions, using *How long* or *How far*.

1 An hour.   5 Two weeks.
2 500 kilometres.   6 It's three kilometres away.
3 Half an hour.   7 Ten hours.
4 20 kilometres.   8 It's a ten-minute drive.

1 *An hour.*   *How long does it take to get from*
                      Buenos Aires to Cordoba by plane?

2 *500 kilometres.* *How far is it from Istanbul to Izmir?*

3 Think of a town in your country. Write sentences saying how far it is from where you are now and how long it takes to get there by different means of transport or on foot.

4 Write six questions about speed and prices in your country.
How fast can you drive in town?
How much does it cost to fly from Buenos Aires to Cordoba?

SPEAKING

1 Compare the information about America in *Vocabulary and listening* activity 5 with your country.
It takes less time to cross my country.
Petrol is more expensive in my country.

2 Think about a memorable journey by car across your country. What is the best route to take?

Now work in pairs and talk about your journeys.

3 Look at some more information about America.

Tallest building	Sears Tower, Chicago, 443 m
Highest mountain	Mount McKinley, Alaska, 6 194 m
Longest river	Mississippi River, 6 020 km
Largest lake	Lake Superior, 82 260 sq km
Biggest city	New York City, population 7.5 million
Hottest place	Death Valley, average temperature 50°C in July
Coldest region	Alaska, average temperatures -12°C in January

Write two questions for each piece of information.
What's the tallest building? How tall is it?

4 Think about answers to the questions about your country. It doesn't matter if you don't know the exact figures.

Now work in pairs and ask and answer questions about your country.
Monique, what's the tallest building in France?
It's the Eiffel Tower.
How tall is it?
It's over three hundred metres high.

Progress check 16-20

VOCABULARY

1 Look at these international words.

> pizza rioja sushi disco rock television football restaurant concert cinema tennis theatre film ballet opera jazz stadium museum video

Put them in these groups: *food, sport, places, types of entertainment, music.*

If you know any more international words, add them to your groups of words.

2 Look at the endings used for these adjectives.

friend**ly** gener**ous** act**ive** dynam**ic** confid**ent** thought**ful** temperament**al** nois**y** cap**able**

Now look at the adjectives in Lesson 18 again. Are there any with similar endings? Words with these endings are often adjectives.

3 Some words are used either for men or for women, but not both. Put a cross (✗) by the sentences which sound odd.

1 He's got a really pretty face.
2 She bought some patterned tights yesterday.
3 He wears a white blouse in the office.
4 She's a very handsome little girl.
5 He's got a warm night-dress for winter nights.
6 She has two pairs of blue jeans.

4 Decide if these items of clothing are usually worn by men or women, or both.

skirt bikini bra knickers underpants tights boots swimsuit shirt shorts jacket sandals pyjamas

5 Look at the vocabulary boxes in lessons 16 – 20 again. Choose words which are useful to you and group them under headings of your choice in your *Wordbank.*

GRAMMAR

1 Complete these sentences with *in, at* or *on.*

1 The football season starts ___ August and finishes ___ May.
2 The ballet is ___ the Apollo Theatre.
3 It's ___ Monday 22 June ___ 7.30pm.
4 The match is ___ 7.15pm ___ Saturday.
5 The Olympic Games ___ 1996 are ___ Atlanta ___ the USA.
6 The film starts ___ 3pm ___ Saturday.

2 Complete these sentences with *to* or *at.*

1 I like going ___ the theatre.
2 I'm working ___ home tomorrow.
3 Shall we meet ___ the cinema?
4 The football match is ___ the main stadium.
5 Would you like to take us ___ the museum?
6 Let's walk ___ the swimming pool.

3 Write questions about Frank.

Ask about:

family likeness age height colour of hair colour of eyes looks

4 Write sentences saying what Frank looks like.

Family likeness: father
Age: 24
Height: 1m 78

48

5 Write sentences about what you look like. Use the questions you wrote in 3 to help you.

6 Choose the best words and complete the sentences.

1 She's got a very ___ face.
   a curly   b tall   c pleasant
2 He's got no hair. He's quite ___ .
   a bald   b fair   c grey
3 He has a ___ grey beard.
   a round   b long   c square
4 He's over two metres. He's quite ___ .
   a short   b honest   c tall
5 He works as a model. He's very ___ .
   a good-looking   b careless   c bossy
6 It's cold today. I'll wear a ___ .
   a swimsuit   b T-shirt   c coat

7 Write the comparative and superlative forms of these adjectives.

big calm careful clever confident friendly generous imaginative informal lazy nervous quiet small smart thoughtful tidy warm

8 Disagree with these statements using the adjective in brackets.

1 Clothes are cheaper than food. (expensive)
2 Mike is taller than Philip. (short)
3 Britain is hotter than Brazil. (cold)
4 It's easier to buy nice clothes in the winter. (difficult)
5 Kate is younger than Penny. (old)
6 The British are more formal than the Germans. (casual)
7 Peter is more polite than Jack. (rude)
8 Graham is lazier than Joe. (hard-working)

1 No they aren't. They're more expensive.

9 Agree with these statements using the superlative form of the adjective.

1 She's very kind.
2 It's a very beautiful town.
3 He's very polite.
4 She's very short.
5 This dress is very expensive.
6 It's a very powerful car.
7 He's extremely handsome.
8 She's very sensitive.

1 Yes, she's the kindest person I know.

10 Complete the dialogue.

A How ___ is it to the nearest station?
B It's two kilometres ___ .
A How ___ does it take ___ car?
B It's ___ five-minute drive.
A How long does it ___ to walk?
B It's thirty minutes ___ foot.
A How ___ is a ticket to London?
B It ___ £5.50.

SOUNDS

1 Say these words aloud.

bl<u>ue</u> g<u>oo</u>d b<u>oo</u>k sh<u>oe</u> b<u>oo</u>t s<u>ui</u>t p<u>u</u>llover t<u>oo</u>k c<u>oo</u>k

Is the underlined sound /ʊ/ or /uː/? Put the words in two columns.

Listen and check.

2 Say these words aloud. Underline the /dʒ/ sound.

geography journalist soldier engineer teenager job manager

Now listen and check.

3 Put these sentences in the correct order and make a dialogue.

a Well, I think I'll leave it. Thank you.
b I'm sorry. This is the largest size we've got.
c Can I help you?
d It's too small. Have you got it in a bigger size?
e Yes, I'm looking for a sweater.
f How about this one? It suits you.

Listen and check. Do you think the speakers sound polite and friendly?

Now work in pairs and say the sentences aloud. Try to sound polite and friendly.

SPEAKING AND WRITING

Work in pairs. You're going to recreate a story called *The Phantom of the Opera and the empty seat.*

Student A: Turn to Communication activity 16 on page 100.
Student B: Turn to Communication activity 3 on page 98.

Present perfect simple (1) for experiences

How are you keeping?

When someone says 'How are you?' do you reply 'Fine thanks, how are you?' or do you say 'I'm not feeling very well. I've had a bad cold, I've been off work, and now I've got a dreadful cough.'? Some people never seem to be ill, others have always got something wrong with them... or think they have.

Try the questionnaire and find out how you're keeping.

		Yes	No
1	Have you ever broken an arm or a leg?	☐	☐
2	Have you ever stayed at home because of illness?	☐	☐
3	Have you ever taken vitamin pills?	☐	☐
4	Have you ever given up any of the following because of your health? smoking	☐	☐
	drinking	☐	☐
	coffee	☐	☐
	meat	☐	☐
	sunbathing	☐	☐
5	Have you ever taken up any of the following because of your health? running	☐	☐
	swimming	☐	☐
	regular exercise	☐	☐
6	Have you ever had an accident while watching a sport?	☐	☐
7	Have you ever had an accident while playing a sport?	☐	☐
8	Have you ever had...? a heart attack	☐	☐
	high blood pressure	☐	☐
	malaria	☐	☐
9	Have you ever had...? flu	☐	☐
	a headache	☐	☐
	food poisoning	☐	☐
10	Have you ever become ill on holiday?	☐	☐
11	Have you ever worried about getting ill?	☐	☐
12	Have you ever stayed in hospital?	☐	☐
13	Have you ever looked up an illness in a medical dictionary?	☐	☐
14	How are you keeping? Not so good.	☐	☐
	I've never felt better!	☐	☐

Mostly **Yes**: Either you've been unlucky with your health or you've become a hypochondriac. Relax! Life's too short to worry so much about your health.

Mostly **No**: You're very lucky ... so far. You're healthy and you don't worry much. But maybe you need to take better care of yourself – just in case.

READING

1 Work in pairs. When was the last time you were ill? Do you worry about staying well? Do you think you're fairly healthy?

2 How are you keeping? Read the questionnaire and find out.

VOCABULARY

1 Complete the diagrams with words for parts of the body.

> tooth mouth eye shoulder
> finger waist knee ankle
> toe foot throat neck
> wrist thumb elbow back

2 Can you name the other parts of the body?

3 Look at this list of parts of the body. Which part doesn't belong?

finger ankle thumb
wrist elbow

'Ankle' is a part of the leg, the others are parts of the arm.

Write some more lists of parts of the body with one part which doesn't belong.

4 Group all the words under four headings: *head, body, arm,* and *leg*.

GRAMMAR

Present perfect simple (1) for experiences

You use the present perfect simple to talk about an action which happened at an indefinite time in the past. You often use it to talk about experiences with *ever* **and** *never.*

Have you *ever stayed* in hospital? (= Do you have any experience of staying in hospitals?)

Yes, I have. (= Yes, at some time in my life, but it's not important when.)

No, I haven't. I've never stayed in hospital.

Remember that if you ask for and give more information about these experiences, such as *when, how, why* **and** *how long,* **you use the past simple.**

When did you *stay* in hospital? *In 1975.*

You form the present perfect simple with *has/have +* **past participle. You usually use the contracted form** *'ve* **or** *'s.*

I've been ill. *He's broken* his leg.

Negative

I haven't had malaria. *I've never had* malaria.

Question

Have you *ever broken* your leg?

9 3 head

5 11 10 6

4 *arm*

7

2 *leg* 12

8 1

1 Rewrite these questions with the present perfect or the past simple of the verb in brackets.

1 (be) you ever in an ambulance?
2 When (be) the last time you (be) ill?
3 (eat) you ever *sauerkraut?*
4 (meet) you ever a famous person?
5 (play) you ever tennis?
6 What (have) you for dinner last night?

2 Write the past participle of these verbs.

be have say eat play teach visit live see work love know pay break meet make try win wear sell

3 Work in pairs. Ask and answer the questions in *How are you keeping?* If you or your partner answer yes, ask for or give extra information.

Have you ever broken your arm or leg? No.
Have you ever been ill on holiday? Yes.
When was that? When we went to India in 1987.

SPEAKING

1 Look at the verbs in *Grammar* activity 2 and think of questions to ask people about their experiences. Write down seven or eight questions.

Have you ever met anyone famous?

2 Find out about the experiences of other people in your class using the questions you wrote in 1. Ask for and give extra information.

Have you ever met anyone famous?
Yes, I have. I've met the President.
Really? When did you meet her?
In 1987.

hair

nose

15 13

face *ear*

16 14

22 | *What's new with you?*

Present perfect simple (2) for past actions with present results

VOCABULARY AND LISTENING

1 You are going to hear Heide Meyer, a teacher in Berlin, talking about what's new with her. She's talking about changes in Germany and in her life since 1989. Which of these words do you expect to hear?

election war government standard of living
employment inflation finance police law and order
freedom rights minister housing tourism education

2 Look at these questions about changes in Germany. Do you think Heide will answer *yes* or *no*?

1 Has there been a change of government?
2 Has the standard of living got better?
3 Have more people got jobs?
4 Has inflation gone down?
5 Have more tourists visited the country?

🔊 Now listen to Heide. Did you guess correctly?

3 🔊 Listen to Heide talking about changes in her own life. Tick (✓) the things that she mentions.

move to a new flat	☐	learn French	☐
finish her studies	☐	find a new job	☐
buy a new car	☐	lose weight	☐
have a baby	☐	stop smoking	☐
learn to cook	☐	get married	☐

4 Match the pairs of sentences about Heide.

1 She's had a baby.
2 She's got married.
3 They've moved to a new flat.
4 She's found a new job.
5 She's bought a new car.
6 She's given up smoking.

a She drives a Volkswagen now.
b It's in a very new building in Potsdam.
c She starts on Monday.
d She feels much healthier.
e She lives with her husband now.
f Their son is one month old.

🔊 Listen again and check.

SPEAKING

1 What's new with you? Think about your life a year or two ago and your life today. Can you think of anything that's different? Make notes.

last year	*today*
lived in London	*live in Bristol*
smoked twenty cigarettes a day	*don't smoke*

2 Work in pairs and talk about what's new with you.

GRAMMAR

> **Present perfect simple (2) for past actions with present results**
>
> **You use the present perfect simple to talk about a past action which has a result in the present. It is not important when the action happened. You often use it to describe changes.**
>
> *She's got married.*
>
> **(She wasn't married before. We don't know when she got married. She's married now.)**
>
> **You often use *just* to emphasise that something has happened very recently.**
>
> *She's **just** had a baby.*
> *They've **just** moved to a new flat.*

1 Look at these past participles.

moved   changed   left   finished
found   got   voted   bought   had
drunk   read   rung   written

Write down their infinitive and past simple forms.

move   moved   moved

Which past participle forms are the same as their past simple forms?

2 Write the verb in brackets in the present perfect form.

1 I (buy) a new car and now I drive to work.
2 He (move) house and now lives in London.
3 She (finish) her book and is now watching TV.
4 We (drink) the whole bottle. It's empty.
5 I (found) my bag. It's here.
6 She (write) me a letter and I'm reading it now.

3 Work in pairs. Check your answers to *Vocabulary and listening* activity 3. Ask and answer questions about Heide.
Has she moved to a new flat?
Yes, she has.

SOUNDS

1 Listen to these sentences. Notice how you don't always hear *'ve* or *'s*.

1 She's stayed late.   She stayed late.
2 They've finished their work.   They finished their work.
3 We've visited America.   We visited America.
4 He's studied French.   He studied French.

2 Now look at the sentences in context and underline the correct verb form.

1 *She's stayed/She stayed* late last Friday.
2 *They've finished/They finished* their work and now they're going home.
3 *We've visited/We visited* America in 1983.
4 *He's studied/He studied* French five years ago.

 Listen and check. Say the sentences aloud.

WRITING

1 Read this letter from Heide to her friend in America and underline any verbs which you can use to describe changes in your life during the last year.

> 1 December
>
> Dear Stacey,
>
> How are you? I've got a lot of news for you. Things have changed quite a lot since I last wrote to you. I've got married, and I now live with my husband. We've moved to a different part of Berlin and we now live in Potsdam. I've stopped smoking and I feel much healthier now. I've found a new job and I can work from home – which is good because the most important news is I've just had a baby. He's one month old and very beautiful.
>
> Here are some recent photos. It would be nice to hear from you.
>
> Best wishes,
>
> *Heide*

2 Look at the notes you made in *Speaking* activity 1. Write full sentences which describe the changes in your life. Use the present perfect.
I've moved to Bristol. I've stopped smoking.

3 Write down some extra information about how your life is now.
I've moved to Bristol. I like it here.

4 Join the two sentences with *and*.
I've moved to Bristol and I like it here very much.

5 Now write a letter to a friend describing how things have changed for you in the last year. Use Heide's letter to help you.

Present perfect simple (3): *for* and *since*

VOCABULARY AND LISTENING

1 Work in pairs. Make a list of important events or festivals in your town or country. (Try not to choose religious occasions.)
Carnival, the Palio, April Fool's Day

2 Look at the words in the box. Which of these words can you use to describe what happens?

> barbecue celebrate dancing
> fair fireworks flag holiday
> king meal parade party
> picnic president prime minister
> queen race soldier speech

Are there any other words you can use?
music, bands…

3 Work in two groups. You are going to hear Barry, who is Australian, talking about an important event in Australia.

Group A: Turn to Communication activity 5 on page 98.
Group B: Turn to Communication activity 18 on page 101.

4 Now work with someone from another group and answer the questions in the *Australia* column.

	Australia	Your country
What's it called?		
When does it take place?		
Where does it take place?		
When did it first take place?		
What happens?		
Is it a public holiday?		
Are there any other interesting features?		

5 🔲 Listen again and check your answers to 4. Can you add any extra information?

6 Choose an important day in your country and complete the column *Your country*.

GRAMMAR

> **Present perfect simple (3): *for* and *since***
> You use the present perfect to talk about an action or state which began in the past and continues to the present.
> You use *for* to talk about the length of time.
> *The 1st October **has been** our national day **for a hundred years**.*
> You use *since* to say when the action or state began.
> *We've had a President **since 1888**.*

1 Complete these sentences with *for* or *since*.

1 They've had a parade ___ 1988.
2 There have been fireworks ___ ten years.
3 It's been a national holiday ___ 1993.
4 The president has given a speech every year ___ five years.
5 She's been Queen ___ March.
6 We've celebrated Independence Day ___ ten years.

2 On a piece of paper, write down something or someone you:

– know – like – want – need
– hate – own – believe

know – Karen, like – sport...

3 Work in pairs and show the list of people and things you wrote in 2. Ask and say how long.
How long have you known Karen?
Since January.
How long have you liked sport?
For ten years.

4 Look at the list of important days or occasions you made in *Vocabulary and Listening* activity 1. Say how long they have been important days or occasions. Use *for*.
Independence Day has been an important occasion for 20 years.

Now say since when your country has celebrated them. Use *since*.
We've celebrated Independence Day since 1864.

SOUNDS

Listen and repeat these sentences. Pronounce for /fə/.

1 They've celebrated it for seventy years.
2 She's lived here for six months.
3 They've had fireworks on Bastille Day for many years.
4 He's worked at the bank for ten years.

WRITING

1 Read this description of an important occasion in Britain, Guy Fawkes Night. Find the answers to these questions.

a How long has it been important?
b What event does it celebrate?
c What happens during the celebration these days?
d What do people have to eat or drink?

We've celebrated Guy Fawkes Night on November 5th every year since 1605. Guy Fawkes was an English Catholic who wanted to kill the Protestant King James I. He tried to destroy the Houses of Parliament in London with a bomb. The plan failed when one of the Catholics warned a relative not to go to Parliament that day. Soldiers arrested Guy Fawkes and he was executed. These days we celebrate Guy Fawkes night with fireworks and a large fire in gardens or in parks. The celebration starts when it's dark. On the fire we put a 'guy', which is made with old clothes and straw and looks like Guy Fawkes. Then the fireworks begin and we burn the guy on the fire. There's usually warm food like sausages and potatoes baked in the fire. I don't know if we celebrate Guy Fawkes because he failed or because he very nearly succeeded.

2 Write sentences describing an important national occasion in your country. Use the questions above and your notes in the chart in *Vocabulary and listening* activity 6 to help you.

We've celebrated Bastille Day since 1789...

Defining relative clauses: *who, which/that* and *where*

Divided by a common language?

George Bernard Shaw said that America and Britain were two nations divided by a common language. But how different is British English from American English? Some British and American people gave their definitions for some common words.

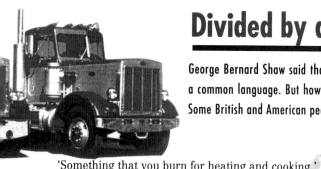

British		American
'Something that you burn for heating and cooking.'	*gas*	'Something you put in your car to make it go.'
'A school which is private.'	*public school*	'A school that is open to everyone.'
'A path which passes under a road.'	*subway*	'A railway which runs under the ground.'
'Something that you wear under your trousers.'	*pants*	'Something which you wear to cover your legs, over your underpants.'
'Clothing which you wear under your shirt.'	*vest*	'Clothing that you wear over your shirt and under your jacket.'
'A list of things that you have bought or eaten in a restaurant and which tells you how much to pay.'	*bill*	'Money which is made of paper.'
'Long sticks of potato which you cook in deep oil and eat hot with a meal.'	*chips*	'Very thin slices of fried potato which you eat cold before a meal or as a snack.'

Confused? British and American English have lots of words which look the same but have different meanings. Nobody ever gets into serious trouble if they make a mistake, although you may get a strange look if you ask for the wrong clothes. But things get even more complicated! Here are some American English words which the British don't use at all.

druggist	someone who sells medicine in a shop.
parking lot	a place where you park the car.
drugstore	a shop where you can buy medicine, beauty products, school supplies, small things to eat.
main street	the street in a town where all the shops are.
stop lights	lights which control the traffic.
faucet	something you turn on and off to control water in a bath or a basin.
elevator	a device which carries people from one floor to another in a building.

But most of the differences between British English and American English are minor and are only concerned with vocabulary, spelling and pronunciation. You can usually understand what words mean from the context. Good luck (British English) or break a leg (American English)!

SPEAKING AND READING

1 Which is more useful to you – British English, American English or another type of English? Work in pairs and say why.

2 Read *Divided by a common language* and decide if these statements are true or false according to the passage.

1 British English and American English are two very different languages.
2 Some words have different meanings in British and American English.
3 Some words are used in one type of English but not in the other.
4 If you don't understand a word, you can usually guess the meaning.

3 Which of the words in the passage did you recognise or know already?

GRAMMAR

> Defining relative clauses: *who, which/that* and *where*
>
> **You use a defining relative clause to define people, things and places.**
>
> **You use *who* for people.**
> *A druggist is someone **who** sells medicine in a shop.*
>
> **You use *which* for things.**
> *A subway is a railway **which** runs under the ground.*
>
> **You often use *that* instead of *who* or *which*.**
> *A druggist is someone **that** sells medicine in a shop.*
> *A subway is a railway **that** runs under the ground.*
>
> **You use *where* for places.**
> *A parking lot is a place **where** you park your car.*

1 Complete the sentences with *who, which/that* or *where*.

1 A hat is something ___ you wear on your head.
2 A post office is a place ___ you can buy stamps and post letters.
3 A journalist is someone ___ writes for a newspaper.
4 A swimsuit is something ___ you wear when you go swimming.
5 A disco is a place ___ you go to dance.
6 A hairbrush is something ___ you use to brush your hair.

2 Write definitions for these words.

kitchen supermarket language teacher fridge pilot
shoes restaurant pub

SOUNDS

1 🔲 Listen to the first paragraph from the passage, spoken first by an American and then by a British person.

Now listen to the second paragraph. Is the speaker American or British?

2 The sound of some words in American English is different from British English.

🔲 Look at the underlined sound and listen to some of the main differences.

f<u>a</u>ther m<u>o</u>ther grandf<u>a</u>th<u>er</u> sist<u>er</u>
<u>o</u>pera c<u>o</u>st c<u>o</u>ncert f<u>o</u>g
bo<u>tt</u>le thir<u>t</u>y dir<u>t</u>y bu<u>tt</u>er
d<u>a</u>nce b<u>a</u>th c<u>a</u>n't pl<u>a</u>nt
n<u>ew</u> T<u>u</u>esday n<u>u</u>clear t<u>u</u>ne

VOCABULARY AND LISTENING

1 Match the American English words with the British English words in the boxes below.

> bill chips druggist faucet french fries gas
> main street vest pants stop lights parking lot
> subway public school

> trousers car park state school traffic lights
> underground waistcoat chemist chips crisps
> high street bank note petrol tap

Check the passage to remind yourself which American English words in the box have different meanings in British English.

2 🔲 Listen to a British person and an American talking about the words and check your answers to 1.

3 Put the words in 1 into categories of your choice. Use a dictionary to find out if there are any other British and American words which can go in each category.

25 *What's it called in English?*

Describing things when you don't know the word

VOCABULARY AND SPEAKING

1 Choose five or six of the words in the box and think of something you can describe with each one.

> cloth cotton curved glass hard heavy high
> leather light long low metal narrow nylon
> oblong oval paper plastic round rubber short
> soft square stone wide wood wool

cotton: a shirt   oval: egg

2 Work in pairs. Tell your partner one of the things you chose. He/she must guess the word to describe it.

> *A: Shirt.*   *B: Cotton?*
> *A: That's right.*

3 Work in pairs. Decide which questions the words in the vocabulary box go with.

> – What does it feel like?   – What's it made of?
> – What shape is it?   – What size is it?

4 Look at some words you can use to describe something if you don't know the English word.

> liquid machine powder stuff thing tool

Think of things you can describe with these words.
toothpaste: stuff

5 Match the objects in the pictures with their names in the box below.

> wallet vacuum cleaner glue towel matches string
> sun-tan lotion soap tin opener sponge bag sweater
> hairdryer camera window cleaner shoe polish
> wine glass pad of paper

LISTENING

1 Look at the objects in the pictures. Think of words to describe what they look like, what they're made of, what shape they are etc.

2 [cassette icon] Listen to people describing five of the objects in the pictures, which they don't know the English word for. Write the name of the object each speaker describes.

Speaker 1 _____   Speaker 2 _____   Speaker 3 _____
Speaker 4 _____   Speaker 5 _____

3 Read these descriptions of some more objects in the pictures. Decide which object each one describes.

a It's a thing you wear to keep warm.
b It's for polishing shoes.
c It's a machine to dry your hair with.
d You use it to take pictures with.
e It's stuff to clean windows with.
f It's for carrying shopping in.
g It's for drinking wine out of.
h It's something to write on.

FUNCTIONS

> **Describing things when you don't know the word**
> *It's a thing you wear **to keep** warm.*
> *It's **for polishing** shoes.*
> *It's **for carrying** shopping **in**.*
> *It's a machine **to dry** your hair **with**.*
> ***You use** it to take pictures **with**.*
> *It's **stuff to** clean windows **with**.*
> *It's **for drinking** wine **out of**.*
> *It's **something** to write **on**.*
>
> **You can use *to* + infinitive to describe the purpose of something.**
> *You use a cassette player **to play** cassettes.*
>
> **You can also use *(to be) for* + *-ing* to describe the purpose of something.**
> *A cassette player **is for playing** cassettes.*

1 Complete these sentences with *in*, *out of*, *with* and *on*.

1 A chair is a thing to sit ＿＿ .
2 A mug is for drinking ＿＿ .
3 A frying pan is for cooking things ＿＿ .
4 A towel is for drying yourself ＿＿ .
5 An oven is for cooking things ＿＿ .
6 A plant pot is a thing for putting plants ＿＿ .

2 Look at the five pictures of objects which have not yet been described. Write sentences describing them using as many expressions from this lesson as possible.

3 Describe the purpose of these things.

1	a sleeping bag	6	a pen
2	an envelope	7	a knife
3	washing-up powder	8	a tea pot
4	a dishwasher	9	a telephone
5	a fridge	10	a typewriter

SOUNDS

1 Say these words aloud. Make sure you pronounce the underlined consonant groups carefully.

p<u>r</u>actical t<u>r</u>ansport news<u>p</u>aper car<u>db</u>oard
too<u>thp</u>aste di<u>shw</u>asher cal<u>c</u>ulator

🔲 Now listen and check.

2 🔲 Listen and notice the words the speaker links.

1 Is it heavy?
2 An egg is oval.
3 A fridge is cold inside.
4 It's for opening tins.
5 It's something to write on.
6 To open, pull it out.

Now say the sentences aloud.

SPEAKING

1 Work in pairs. There are several ways to refer to something if you don't know the English word. You can describe it, draw it or mime it. You can also point if you can see one.

Look at the objects in the pictures. Which is the best way to refer to them?
I think I would draw a wine glass.

2 Work in groups of three or four. You're going to play *What's it called in English?* Here are the rules.

> # What's it called in English?
>
> **AIM:** for one person in each group to refer to a word by describing it, drawing it, miming it or pointing to it. The first group to guess all the words is the winner.
>
> *1* One student from each group comes to the teacher who gives them a word. They must describe it, draw it, mime it or point to it. They mustn't say the word!
>
> *2* The student who guesses the word correctly goes to the teacher and tells him/her what the word is. If it is correct, the teacher gives him/her another word.
>
> *3* The student acts out the word, and the game continues until a group has guessed all the words. The first group to finish is the winner.

Progress check *21-25*

VOCABULARY

1 Write the verbs which come from these nouns. Underline the suffixes for each noun.

government employment election education information reading writing

*govern govern**ment***

Now write nouns which come from these verbs, using the suffixes you underlined.

entertain refresh teach greet exhibit connect manage

When you write new nouns in your *Wordbank* you may like to group them according to their suffix.

2 *Get* is a common verb in spoken English. Here are some of its meanings.

1 *get* + past participle = *become*	*They've got married.*
2 *get* + adjective = *become*	*I'm getting angry.*
3 *get* + object = *fetch, obtain*	*Could I get you a drink?*
4 *get* + object = *catch, take*	*I get a train at 7.15.*
5 *get* + object = *receive*	*We got the news yesterday.*
6 *get* + to + object = *arrive in/at*	*I got to work at 8am.*
7 *get* + preposition has the idea of movement, sometimes with difficulty.	*We get up at 7.30.*
	He got in through the window.

Look at these sentences and decide which meaning of *get* they show. Rewrite them replacing *get* with another verb.

a It's getting dark.
b We got a call from her.
c The train gets to London at 6 o'clock.
d He got the tickets at the box office.
e The child got down from the table.
f She gets worried very easily.
g The bus to get is the number 39.

It's a good idea to note the different meanings of *get* as you come across them.

3 Look at the vocabulary boxes in lessons 21 – 25 again. Choose words which are useful to you and group them under headings of your choice in your *Wordbank*.

GRAMMAR

1 Write the past participles of these verbs.

be become break bring come do find get go have learn lose put read say sit sleep take think understand write

2 Write short answers to these questions.

Have you ever...

1 ...visited the USA?
2 ...learnt a musical instrument?
3 ...been on TV?
4 ...met a rock star?
5 ...tasted English beer?
6 ...eaten English cheese?
7 ...heard Beethoven's Fifth Symphony?
8 ...written a letter to a newspaper?

3 You haven't seen your friend Jenny for five years. Ask her questions using the words below.

1 find a new job
2 get married
3 stop smoking
4 buy a motorcycle
5 visit USA
6 write a book
7 travel around Europe
8 learn to cook

1 Have you found a new job?

4 Here are Jenny's answers to the questions in 3. Write full answers.

1 Yes.   2 No.   3 Yes.   4 No.
5 Yes.   6 No.   7 Yes.   8 No.

1 Yes. She's found a new job.

5 Complete these sentences with the present perfect or the past simple form of the verb in brackets.

1 I never (drive) a car in my life.
2 I (leave) my secondary school in 1989.
3 We not (meet) before. My name's John.
4 Where you (buy) your coat?
5 What you (pay) for it?
6 How long you (be) here now?
7 Last Sunday I (get up) at eleven.
8 You (see) the latest Harrison Ford film?

6 Answer the questions. Use *for* and *since* in turn.

How long have you...

1 ...been a student of English?
2 ...lived in your home?
3 ...known your best friend?
4 ...liked your favourite food?
5 ...been in class?
6 ...had the clothes you're wearing?

7 Say what you use these things to do.

1 an umbrella   4 an overcoat
2 sunglasses   5 a saw
3 sticky tape   6 water

8 Say what these things are for.

1 a garage   4 a toothbrush
2 a bath   5 a door handle
3 a wardrobe   6 a kettle

SOUNDS

1 Say these words aloud.

surfing fireworks primary learnt economics
lecture heard university chosen word

Is the underlined sound /ɜ:/ or /ə/? Put the words in two columns.

🔊   Listen and check.

2 Say these words aloud.

sport up awful walking summer bus pub
sunny performance four

Is the underlined sound /ɔ:/ or /ʌ/? Put the words in two columns.

🔊   Now listen and check.

SPEAKING AND WRITING

1 Look at these experiences.

– visit the Great Wall of China
– play tennis
– drive a Porsche
– cook dinner for six friends
– swim in a river
– see a play by Shakespeare
– write a diary
– read *Time* magazine

Think of two more experiences and add them to the list.

2 Now find people in the class who have done the things in 1. Ask two extra questions about these experiences.
Have you ever visited the Great Wall of China?
Yes, I have.
Really! When was that?   In 1989, when I was twelve.
What was it like?   Fantastic!

3 Write a paragraph describing what you have learnt about the other students.
Two people have visited the Great Wall of China. Paco went in 1989, when he was twelve...

Safety first

Modal verbs; *must* for obligation; *mustn't* for prohibition

SPEAKING

Work in pairs and look at these safety instructions.
Who do you think is speaking and where?

1 'You must fasten your seat belt.'
2 'You mustn't lean out of the window.'
3 'You must put out your cigarette.'
4 'You mustn't leave them on their own.'
5 'You must wear strong shoes.'
6 'You mustn't play with matches.'
7 'You must stay in your seat until we stop.'
8 'You must wear a helmet.'

GRAMMAR

> **Modal verbs**
>
> **Modal verbs:**
> – have the same form for all persons
> – don't take the auxiliary *do*
> – take an infinitive without *to*.
>
> *Must* for obligation, *mustn't* for prohibition
> ***Must*** is a modal verb. You use ***must*** to talk about
> something you're obliged or strongly advised to do.
> You often use it when you talk about safety
> instructions.
> *You **must** fasten your seat belt.*
>
> You use ***mustn't*** to talk about something you aren't
> allowed to do or you're strongly advised not to do.
> *You **mustn't** lean out of the window.*
>
> For strong prohibition you use ***must never***.
> *You **must never** walk on the railway line.*
>
> ***Must*** and ***have to*** have almost the same meaning.
> You usually use ***must*** when the obligation comes
> from one of the speakers.
> *I usually forget her birthday. I **must** remember this year.*
> *The baby's asleep. You **must** be quiet.*
>
> You usually use ***have to*** when the obligation comes
> from a third person. You often use it when you talk
> about rules.
> *The government says you **have to** do military service.*
> *You **have to** show a cheque card when you pay by
> cheque.*

1 Think of five or six things you must do to make your life safer.
   I must stop smoking.
   I must mend the brakes on my car.

2 Complete these sentences with *must* or *mustn't*.

   1 You ___ eat less to lose weight.
   2 You ___ wear a seat belt while the plane is taking off.
   3 You ___ travel on a train without a ticket.
   4 You ___ ride a bicycle on a motorway.
   5 You ___ always lock your car.
   6 You ___ drive at more than 55 km/h in town.

3 Think of other rules or safety instructions for these situations.

   – in a plane   – in a car   – in the street   – on a boat
   – in the mountains   – in the desert

   You mustn't smoke during take-off and landing.

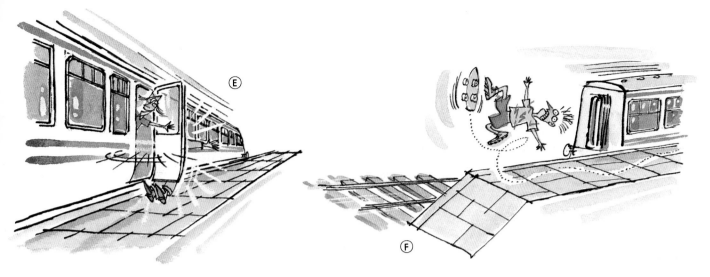

(E)

(F)

READING

1 Look at the drawings for a guide to railway safety. Match them with the rules below.

1 Do not ride a bicycle on a station platform. Do not use a skateboard there either.
2 Never, never, never stick your head out of a window of a moving train.
3 Do not throw anything out of the windows.
4 Never open a door before the train has stopped.
5 Never go onto a railway line or walk along it.
6 Never put anything on the railway line.

2 Work in pairs and say which drawing you like best.

3 Rewrite the sentences in 1 saying what you *mustn't* do.
1 You mustn't ride a bicycle on a station platform.

SOUNDS

1 Say these words.

a /mʌs/ b /mʌst/ c /mʌsn/ d /mʌsnt/

🔊 Listen to these sentences and decide if you hear *a, b, c* or *d.*

1 You must go now.
2 You mustn't be late.
3 Yes, you must.
4 No, you mustn't.
5 You must visit us.
6 We must have lunch.
7 He mustn't say that.
8 You mustn't eat much.

Now say the sentences aloud.

2 🔊 Listen to the way speaker *B* uses a strong stress and intonation to sound insistent.

A I don't want to go to school. **B** You must go.
A Must I really go? **B** Yes, you must.
A I'll be late. **B** You mustn't be late.
A Well, I'm going to leave early. **B** No, you mustn't.

Now work in pairs and say the sentences aloud. Try to sound insistent.

VOCABULARY AND LISTENING

1 Match the words in the vocabulary box with the situations below.

– on a motorway   – in a train
– at the border   – in the street

bicycle carriage customs drive
duty-free fast first class guard
lane pavement police officer
ride second class speed
ticket inspector traffic

2 Work in pairs and say what you have to or mustn't do in the situations in 1.

3 🔊 Listen to three conversations and decide what the situation is. Choose from the situations in 1.

4 Work in pairs and say what the people in the conversations have to or mustn't do.
He mustn't drive so fast.

🔊 Listen again and check.

5 Talk about rules or safety instructions you have at your language class or school. How many can you think of?
You mustn't drop litter.
You have to do your homework.

27 | *The Skylight*

Can, could (1) for ability

SPEAKING

1 Find someone in the class who can:

– swim a hundred metres
– speak a foreign language (not English)
– drive a car
– stay awake all day
– stay up all night

– write with their left hand
– ride a bicycle
– use a computer
– cook
– play a musical instrument

Can you swim a hundred metres? Yes, I can. No, I can't.

2 Choose the last person you spoke to and find two things you can both do, and two things neither of you can do.

GRAMMAR

> **Can, could (1) for ability**
>
> **Can** is a modal verb. It has the same form for all persons and you don't use the auxiliary **do** in questions and negatives. You use **can** to express general ability, something you are able to do on most occasions.
> *I **can** swim a hundred metres.*
>
> | **Negative** | **Questions** | **Short answers** |
> |---|---|---|
> | *I can't ride a bicycle.* | *Can you swim?* | *Yes, I can.* |
> | | | *No, I can't.* |
>
> You can also use **be able to** to express general ability. It is more formal than **can** and is mostly used in the future or the past tense.
>
> You use **could** and **couldn't** or **wasn't/weren't able to** in the past to express general ability.
> *When I was five, I **could** swim, but I **couldn't** write my own name.*

1 Think about when you were five years old. Which things in *Speaking* activity 1 could you do? Write sentences.
I could swim, but I couldn't cook.

2 Find out what other people could or couldn't do when they were five years old.

LISTENING AND SPEAKING

1 Work in pairs. You are going to hear a story in four parts, adapted from *The Skylight* by Penelope Mortimer. Here are some key words from the first part. What do you think the story is about? (The words are in the right order.)

taxi woman five-year-old boy sleep house summer mountains hot afraid suitcases shutters doors closed locked

2 🔲 Listen to part 1 of the story and check your answers to 1.

3 Here are some questions about part 2 of the story. Try and guess the answers.

1 Why didn't they go in the house?
2 What did she see on the roof?
3 How did she get up there?
4 Did she get through the skylight?
5 Who climbed through the skylight?
6 What did she ask him to do when he was in the house?
7 Did he do it?

4 🔲 Listen to part 2 of the story and check your answers to the questions in 3.

5 Here's part 3 of the story with some words and phrases missing. Can you guess what they are?

It was now dark. She went down again and ran round the house, shouting his _____ . Something has happened to him, I must go for _____ . She ran to the road and when she saw the lights of a _____ , she waved her arms to _____ it. She started to _____ . It was a long time before the three men understood.

'But how can we get in? We have no tools,' they said.

'There's a _____ back there. Will you take me?' They let her into the car.

'Turn _____ . It's back there on the left. There it is!'

They turned off the _____ She got out of the car and ran to the front _____ .

A small woman in trousers opened the door. 'My dear, what's _____ ?'

'You're English?'

She told her the story. Another woman appeared. 'Yvonne,' Miss Jardine said, 'Get some tools, a hammer and an axe.'

They all got into the _____ and went back to the _____ . They drove up the _____ and stopped. She ran to the house, calling 'Johnny? Johnny?'

6 🔲 Listen to part 3 of the story and check your answers to 5.

7 Work in pairs. What do you think happens at the end of the story?

8 Before you listen to part 4 of the story, check you understand these words.

axe hammer toys smash thumb

🔲 Now listen to part 4 of the story and find out what happens.

VOCABULARY

1 Here are some words from the story. Check you know what they mean. (You can use a dictionary.) Try to remember where they came in the story.

> hill sleep asleep narrow stones grass suitcase
> square skylight roof shutter door lock knock
> ladder lower climb shout hurry wave farm tools
> axe hammer toys smash thumb whisper

2 Can you think of categories to group the words? Think of other words which go with them.

28 | *Breaking the rules?*

Can, can't (2) for permission and prohibition

READING

1 Work in pairs. Look at the signs in the photos. Do you understand what they mean? Where do you think the signs are? Choose from these situations.

– in the street   – in a bar
– in a park   – in a shop
– in a church   – in a cinema
– in a restaurant

2 Read *Breaking the rules?* and answer the questions with *yes* or *no* for your country.

Breaking the rules?

These days there are rules everywhere we go and it's hard to obey them all. But what about the rules and customs we don't know about when we're visitors to a town or a country? Are you breaking the rules?

1 You're waiting to cross the street at a pedestrian crossing. The lights are red for pedestrians, but there are no cars. Can you cross?

2 You're in a large city park which has lots of trees and grass. It's a beautiful summer's day and the sun is shining. Can you walk on the grass?

3 It's half past three on a Sunday afternoon, and you're thirsty. Can you buy a nice, cold glass of beer?

4 You're with some very young children and you'd like to have a drink in a bar. Can they go in with you?

5 You're in a restaurant but you don't know what the food is like. Can you walk into the kitchen and have a look?

6 You want to buy a postcard in the newsagent's or tobacconist's shop. Can you also get stamps there?

7 You're in the cinema and you'd like a cigarette. There are no signs which say 'no smoking'. So can you smoke?

8 You're in a cafe and you'd like a drink. Can you get it at the bar and pay for it there?

9 You feel quite ill, but you don't have any medical insurance or much money. Can you make an appointment to see the doctor?

10 You want to take the bus. Can you buy a single ticket before you get on?

GRAMMAR

> *Can, can't* (2) for permission and prohibition
>
> *Can* is a modal verb. It has the same form for all persons and you don't use the auxiliary *do* in questions and negatives. You use *can* + infinitive to talk about what you're allowed to do or what it is possible to do.
> *You **can** cross when the light is green.*
>
> You use *can't* to talk about what you're not allowed to do or what it is not possible to do.
> *You **can't** cross when the light is red.*
>
> | Questions | Short answers |
> |---|---|
> | ***Can*** *you cross when the light is green?* | *Yes, you **can**.* |
> | ***Can*** *you cross when the light is red?* | *No, you **can't**.* |
>
> *Can't* and *mustn't* mean the same.
> *You **mustn't** drive a car if you're only sixteen.*
> *You **can't** drive a car if you're only sixteen.*

1 Complete these sentences with *can* or *can't* and make true statements about your country.

1 You ___ drive a car when you're sixteen.
2 You ___ go into a bar when you're fourteen.
3 You ___ get married when you're sixteen.
4 You ___ wash your car on Sundays.
5 You ___ visit the USA without a visa.
6 You ___ get cheap housing.

2 Work in pairs and check your answers to the questions in *Breaking the rules?*.
In my country you can't walk on the grass in parks.

SOUNDS

1 Work in pairs. Say these sentences aloud.

'Can I come in?'
'Yes, you can. You can sit down there.'
'Can I smoke?'
'You can't smoke in here, but you can smoke outside.'

Is the underlined sound /æ/, /ə/ or /ɑː/?

🔊 Listen and check.

2 In American English it is sometimes difficult to hear the difference between *can* and *can't*.

🔊 Listen and tick (✓) the word you hear.

1 You *can/can't* smoke.
2 You *can/can't* walk on the grass.
3 You *can/can't* cross now.
4 You *can/can't* go.

LISTENING

1 🔊 Listen to Jane, a British woman, answering the questions in *Breaking the rules?*. Put a tick (✓) if the answer is *yes* and a cross (✗) if the answer is *no*.

2 Work in pairs and check your answers. Are the rules and customs the same as in your country?

🔊 Listen again and check.

VOCABULARY AND SPEAKING

1 Look at the words in the box. Which of the following things can you and can't you do in the situations?

– read a newspaper – wear shoes – wear a hat
– go swimming – sunbathe – watch TV
– go for a walk – cross the road
– ride a motorbike – wear jeans – listen to music
– smoke – eat – drink – go to sleep – talk
– sing – take your dog – feed the animals

> escalator lift hotel bar petrol station church mosque
> hospital level crossing office shop beach mountain
> canal library stadium traffic lights pedestrian crossing
> park museum prison aeroplane bus train zoo

2 Talk about other rules or customs in your country?
In my country, children can go into a bar but they can't drink alcohol.

Warning: flying is bad for your health

Should and shouldn't for advice

READING

1 Look at the title of the passage. Why do you think flying may be bad for your health?

2 Read *Warning: flying is bad for your health* and find out if it is safe for you to fly.

3 Write down the reasons why flying is bad for people's health.
Less oxygen, changes of pressure...

Flying is the safest way to travel ... or is it? Some doctors think the aeroplane is a dangerous place, especially for the old or the unhealthy.

Although the aeroplane is pressurised, there is less oxygen than on the ground. Anyone with heart disease or a lung problem notices the difference much sooner. Even healthy people find it difficult to concentrate after hours of breathing less oxygen than usual. So anyone who has had a heart attack should not fly for at least two weeks after the attack. After an operation, you should stay on the ground for at least ten days.

Because of changes of pressure, pregnant women shouldn't take a flight lasting more than four hours after their thirty-fifth week or a shorter flight after 36 weeks. People with bad colds will probably get earache during take-off and landing.

Even if you feel well when you get on the plane, you will possibly feel ill when you get off. Sitting on a plane for many hours – especially in economy class where there isn't very much leg room – gives everyone aches and pains, so you should take some exercise, especially on long flights.

Most of the air you breathe is recycled so you will possibly catch a cold or flu from one of the other passengers.

Flying also causes dehydration. If you drink or eat too much, you'll wake up feeling ill. Everyone needs to drink more in the air, but you shouldn't drink alcohol because it makes you even more thirsty.

The most common problem is jet lag. But there isn't much you can do to prevent it. You should change to your new time zone as soon as possible and you shouldn't sleep if it's still daylight.

Crowded airports, long queues and delays cause stress and high blood pressure. So, be careful! Flying is the safest way to travel, but is it the healthiest?

Adapted from *The Independent on Sunday*

WARNING: FLYING IS BAD FOR YOUR HEALTH

GRAMMAR

> *Should* and *shouldn't* for advice
>
> *Should* is a modal verb. It has the same form for all persons and you don't use the auxiliary *do* in questions and negatives. You use *should* + infinitive to give advice.
> You *should* take some exercise on long flights.
> You *shouldn't* drink alcohol.
>
> | Questions | Short answers |
> |---|---|
> | *Should* I take some exercise? | *Yes, you* ***should***. |
> | *Should* I drink alcohol? | *No, you* ***shouldn't***. |
>
> *Should(n't)* and *ought(n't) to* mean the same.
> You *ought to* take some exercise on long flights.
> You *oughtn't to* drink alcohol.

1 Give advice to these people. Use these words and phrases.

see a doctor   go to the dentist   take some aspirin
lie down   take some exercise   take some medicine

1 I've got a headache.
2 My arm hurts.
3 I've got toothache.
4 I've got a sore throat.
5 My legs are stiff.
6 I've got a temperature.

2 Look at the article again and give advice to people who travel by air.

People with heart disease should be careful.

VOCABULARY

1 Work in pairs. Say what you should do if you are:

– ill   – sick   – sore   – thirsty   – tired

2 Look at the words and expressions in the box and write them under two headings: *parts of the body* and *complaints*.

ache   arm   high blood pressure   cold   disease   ear   flu
hangover   head   heart   insect bite   jet lag   leg   lung
nose   pain   stomach   sunstroke   temperature   throat

Which words go together to make a complaint?
headache...

LISTENING

1 You are going to hear a doctor talk about what you should and shouldn't do to stay healthy when you travel. Match the complaints with the advice.

complaint	advice
jet lag	Drink lots of water.
	Don't spend all day in the sun.
stomach upsets	Take an insect repellent.
	Cover your arms and legs.
insect bites	Wear a hat.
	Drink only bottled or boiled water.
sunstroke	Don't eat uncooked food.
	Try to stay awake.

2 🔊 Listen to the doctor and check if you were right.

3 Try and remember any other advice the doctor gives.

🔊 Listen again and check.

SPEAKING

Work in pairs. Talk about possible dangers to your health in your country. Think about the following and say what you should or shouldn't do to stay healthy.

– the environment   – the climate   – animals
– the livestyle   – dangerous sports

You shouldn't go skiing on your own when it's snowing heavily.

30 *Doing things the right way*

Asking for permission; asking people to do things; offering

1 **You're a guest in someone's home. You'd like a cigarette. What do you say?**
a 'Is it all right if I smoke?'
b 'Would you like a cigarette?'
c Nothing and light up.

2 **A friend suggests you have dinner together in a certain restaurant. At the end of the meal, the waiter brings the check. What do you say?**
a 'I'll pay this.'
b 'Shall we share this?'
c Nothing. Your friend suggested dinner, and you expect him to pay.

3 **You're visiting a friend when the phone rings. What do you expect her to say to the caller?**
a 'It's great to hear from you. Hold on while I get a chair.'
b 'Would you mind if I called you back? I've got a visitor here at the moment.'
c Nothing. It's rude to pick up the phone when you've got guests.

4 **It's late and your neighbours are playing very loud music. What do you say to them?**
a 'Turn down the music!'
b 'Could you turn the music down, please?'
c Nothing. You call the police.

5 **You meet a Ms Esther Craig for the first time. You don't know how to address her. What do you say?**
a 'What do I call you?'
b 'Can I call you Esther?'
c 'Would you mind telling me what to call you?'

6 **You meet someone at a party and get on very well. As she leaves, she says, 'Nice meeting you. We must do lunch sometime.' What do you say to her?**
a 'Great! Shall we make a date?'
b 'Would you mind giving me your phone number?'
c 'That's a great idea. Bye!'

7 **You'd like your friend to lend you a book. What do you say?**
a 'Lend it to me, will you?'
b 'It's much too expensive for me to buy.'
c 'Would you mind lending it to me?'

8 **You're at the information office at the railway station and you want to know some train times. What do you say?**
a 'When's the next train to New York?'
b 'Tell me when the next train to New York is!'
c 'I wonder if you could tell me when the next train to New York is.'

9 **Your host serves you food you don't like. You eat it, but then the host offers you more. What do you say?**
a 'It was very nice, but no thank you. I've had enough.'
b 'Yes, would you mind if I only had a little?'
c 'No way. What was that, anyway?'

10 **As you're leaving a shop the assistant says, 'Have a nice day!' What do you say?**
a 'Thank you. Same to you. Bye.'
b 'Have a nice day yourself!'
c 'No, thanks. I've made other arrangements.'

more?

READING AND LISTENING

1 Read the questionnaire *Doing things the right way*. What do you do or say in your country?

2 🔲 Listen to two Americans talking about the questionnaire and tick (✓) their answers.

3 Work in pairs. Try to remember what the speakers said about each possible answer.

🔲 Listen again and check.

4 Compare your answers to the questionnaire with the speakers' answers. Do you do things the same way?

FUNCTIONS

Asking for permission	Asking people to do things
Can I smoke?	*Can you speak louder, please?*
Could I leave now?	*Could you help me?*
May I call you Esther?	*Could you tell me where the station is?*
Is it all right if I smoke?	*I wonder if you could help me.*
Would you mind if I borrowed it?	*Would you mind lending it to me?*

Could **is a little more formal than can.** *May* **is very formal.**

Agreeing	Refusing
Yes, of course. Yes, go ahead.	*(I'm sorry,) I'm afraid I can't/ you can't.*
By all means.	*(I'm sorry,) I'm afraid not.*

In response to *Would you mind...?*

No, of course not. No, go ahead.	*(I'm sorry,) I'm afraid I do (mind).*

Offering

	Accepting	Refusing
Shall I do that?	*That's very kind of you.*	*No, it's all right, thank you.*
I'll do that, shall I?	*Thank you.*	*No, I'll do it.*

1 Rewrite these questions using the phrase in brackets.

1 Can I leave early. (Would you mind...?)
2 Can you help me? (Would you mind...?)
3 Where's the station? (I wonder if...?)
4 What's your phone number? (Could you tell me...?)
5 Can I open the window? (Is it all right if...?)

1 Would you mind if I left early?

2 Write suitable responses to the questions in 2.

1 Agree. 2 Agree. 3 Refuse. 4 Agree. 5 Refuse.

1 Yes, of course.

SOUNDS

🔲 Listen to these sentences. Does the speaker sound rude or polite?

1 Is it all right if I smoke?
2 Could you turn the music down, please?
3 Would you mind if I called you back?
4 Would you mind telling me what to call you?
5 Would you mind giving me your phone number?

Now say the sentences aloud. Try to sound polite.

VOCABULARY AND SPEAKING

1 Here are some new words from this lesson. Check you know what they mean. (You can use a dictionary.) Can you remember the sentence where you saw them?

share hold on neighbour turn down get on with someone date phone number lend borrow rude polite formal informal manners

2 Go round the class asking for permission to do things or asking people to do things. Be polite.
Is it all right if I open the window?
Yes, of course.

Progress check 26-30

VOCABULARY

1 Some adjectives and nouns often go together. For example:

high temperature   red light   sore throat

Match the adjectives in box A with the nouns in box B. Some adjectives can go with more than one noun.

A | alcoholic   bad   duty-free   first-class   heavy high-speed   narrow   next-door   steep

B | cigarettes   cold   drink   hill   manners neighbour   path   suitcase   ticket   train

Think of other nouns which the adjectives often go with, and other adjectives which the nouns often go with.

2 Look at these words and find out:

– how many meanings they have
– how many parts of speech they can be.

head   complaint   cold   back   neck   ache   sore   arm

3 Complete these sentences with the words in 2. There are some extra words.

1 He held the bottle by the ___ .
2 I'm very angry. I'd like to make a ___ .
3 I'm tired. Let's go ___ home.
4 I feel ___ . Put the heating on.
5 She was the ___ of the department.
6 My legs ___ .

4 Here are some strategies you can use when you are reading and you come across a word you don't understand.

– Decide what part of speech the word is.
– Decide if the word is important to the general sense of the passage.
– Try and guess the general sense of the word.
– Read on and confirm or revise your guess.

5 It is useful to learn new words by associating them with other words. For example:

sunstroke – temperature, headache, lie down
thirsty – drink, water, hungry

Look at the vocabulary boxes in lessons 26 – 30 again. Choose words which are useful to you and group them under headings of your choice in your *Wordbank*. Then think of two or three words that you associate with each one.

GRAMMAR

1 Complete these sentences with *must* or *mustn't*.

1 You ___ wear a warm coat, because it's cold.
2 You ___ be rude to your teacher.
3 You ___ stand too close to the edge of the platform.
4 You ___ be late for an important meeting.
5 You ___ drive slowly, as I'm very nervous.
6 You ___ make too much noise in a library.
7 You ___ do your homework every day.
8 You ___ feed the animals in the zoo.

2 Write sentences describing three things you must do in your English classes and three things you mustn't do.
You must speak English.

3 Write five things you couldn't do when you were ten years old but you can do now.
I couldn't speak English, but I can now.

4 Say if you can or can't do these things when you fly.

Can you...

1 ...arrive at the airport just before the plane leaves?
2 ...stand up during take-off?
3 ...smoke during landing?
4 ...have a drink during the flight?
5 ...visit the pilot?
6 ...listen to your personal stereo?
7 ...use a mobile phone?
8 ...buy a ticket at the airport?

1 No, you can't.

5 Reply to these people and give advice. Use *should(n't)* and *ought(n't) to* in turn.

1 I feel so tired.
2 I don't get much exercise.
3 I stay in bed all day on Sundays.
4 I've hurt my leg playing football.
5 My wife doesn't understand me.
6 I need to learn English quickly.
7 I don't have many friends.
8 I spend twelve hours a day at work.

1 You should go to bed early tonight.

6 Rewrite these sentences with *can, should, have to* or *can't*.

1 You aren't allowed to park a car on the pavement.
2 You are allowed to cross the road at traffic lights.
3 He ought to wear a tie.
4 You are obliged to pass your driving test.
5 You aren't allowed to drink and drive.
6 She ought to see a doctor.
7 Men are obliged to do military service in many countries.
8 You are allowed to go to night-clubs when you're sixteen.

1 You can't park a car on the pavement.

7 Rewrite these requests more politely.

1 Turn it off!
2 I want to smoke.
3 Help!
4 Where's the police station?
5 I want to borrow your book.
6 Lend it to me!
7 What's the date?
8 Speak up!

SOUNDS

1 Say these words aloud.

wr<u>o</u>te v<u>o</u>te kn<u>ow</u> b<u>oy</u> <u>o</u>nly teleph<u>o</u>ne
h<u>o</u>me n<u>oi</u>se r<u>oy</u>al ph<u>o</u>to unempl<u>oy</u>ment

Is the underlined sound /əʊ/ or /ɔɪ/ ? Put the words in two columns.

🔊 Listen and check.

2 Say these words aloud.

<u>sh</u>oe sta<u>ti</u>on tea<u>ch</u>er pre<u>ss</u>ure oxy<u>g</u>en pa<u>ss</u>engers tempera<u>t</u>ure
situa<u>ti</u>on <u>str</u>anger

Is the underlined sound /ʃ/, /tʃ/ or /dʒ/? Put the words in three columns.

🔊 Listen and check.

3 Say these sentences aloud. Try to sound polite and friendly.

1 Could you turn down the music, please?
2 Could you be quiet please?
3 You must fasten your seat belt.
4 I'm sorry, but you can't smoke in here.
5 Can I come in?
6 Where's the nearest bank?

🔊 Listen and check.

SPEAKING AND WRITING

1 Work in groups of two or three. Write some advice for foreign visitors to your country. Think about the following:

– where to go – where not to go – what to see – what to buy
– what to do – what to wear – what to drink – what to eat

2 Write some rules for foreign visitors to your country. Think about the following.

– what you must be careful of
– what you mustn't bring into the country
– what you must wear in certain places
– what you mustn't photograph
– what you must do when you enter someone's home
– what you mustn't do in the streets

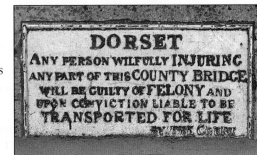

3 Compare your advice and rules with those of other groups. Choose the five most useful pieces of advice and the five most important rules.

My strangest dream

Past continuous (1) for interrupted actions: *when*

LISTENING AND SPEAKING

1 Work in pairs. You're going to hear an English woman talking about her strangest dream, which she called *The day the Queen came to tea*. First of all, think about these questions.

– How does the Queen of England usually travel?
– Do you think she does her own shopping?
– Do you think she ever visits people in their own homes?
– Does she ever do anything which is unexpected?

2 Look at the words below which come from the first part of the dream. What do you think it's about? (The words are in the right order).

sit fire knock front door surprised Queen crown
bag shops come in front room shopping tired
cup tea chocolate biscuits curtain material castle
redecorate burn down

3 🔊 Listen to the first part of the dream and tick (✓) the words as you hear them. Did you guess correctly in 2?

4 Here are some extracts from the second part of the dream. (They are in the wrong order.) What do you think happens next in the dream?

a '... and now she's going home.' ☐
b ...waved to me like she does on television. ☐
c It was Prince Philip. ☐
d Suddenly the phone rang. ☐
e 'Would you like a lift home?' ☐
f ... my husband arrived home from work. ☐
g '...I'll get the bus.' ☐
h and we carried on talking just as if we were old friends... ☐
i 'I'll be back at about five o'clock...' ☐
j 'Hello, Queen. Pleased to meet you.' ☐

5 🔊 Listen to the second part and number the extracts as you hear them. Did you guess correctly in 4?

GRAMMAR

Past continuous (1) for interrupted actions: *when*
You form the past continuous with ***was/were*** + present participle (verb + ***-ing***).

*I **was** watch**ing** television.*
*We **were** talk**ing** about the weather.*

You use the past continuous to talk about something that was in progress at a specific time in the past.

What were you doing at nine o'clock yesterday morning?
I was going to work.

You also use the past continuous tense to talk about something that was in progress at the time something else happened or interrupted it. You join the two parts of the sentence with *when*. The verb in the *when* clause is usually in the past simple.

*I **was watching** television **when** there **was** a knock at the door.*
*We **were talking** about the weather **when** the phone **rang**.*

Remember that you don't usually use these verbs in the continuous tenses:
*believe feel hear know like see smell sound
taste think understand want*

---

1 Join the two parts of the sentence from the dream with *when*.

1 I was sitting in front of the fire...   a my husband arrived home from work.
2 I was getting the tea ready...
3 She was looking for some curtains...   b she stopped and waved.
4 She was finishing her tea...   c she called out to me.
5 She was going down the garden path...   d there was a knock at the door.
   e she saw this lovely material.

2 Work in pairs. Ask and say what you were doing at these times yesterday.

1 7.15am   2 8.35am   3 9.30am   4 1pm   5 5.15am   6 7.45pm

What were you doing at seven fifteen yesterday morning?
I was getting up. What were you doing?
I was having breakfast.

3 Choose the best verb.

1 I *understood/was understanding* what she said.
2 She *had/was having* dinner when the phone *rang/was ringing*.
3 It *rained/was raining* when he *got/was getting* into his car.
4 He *made/was making* me a sandwich because I was hungry.

VOCABULARY AND WRITING

1 In the dream you heard there were a number of verbs followed by a preposition. Match the verbs and the prepositions you heard. Try to remember the sentence you heard them in.

Verbs	be call come get go look pick put show stand turn wave
Prepositions	to on in in front of at for into out up

Are there other verb + preposition combinations above? Choose four or five and use them in sentences.

2 The dream also included some two-word nouns (compound nouns) or an adjective and a noun combination.
tea cup, front door

How many combinations did you hear with these words?

biscuit chocolate cup door front garden path room tea

Are there any other combinations you can make with the words in the box?

3 You often use these words and expressions in stories.

suddenly fortunately
unfortunately finally
to my surprise

Complete the sentences with the words or expressions.

1 We had a wonderful holiday. ___ we had to go back to work.
2 It was late at night and I was asleep. ___ I heard a strange noise.
3 The guests stayed very late. ___ they said goodbye and left.
4 There was a car accident. ___ no one was hurt.
5 I lost my wallet. ___ someone found it and gave it back.

4 Rewrite the dream of *The day the Queen came to tea*. Use the words and extracts in *Listening* activities 2 and 4, and some of the words and expressions in 3 above.
I was at home, watching television. Suddenly there was a knock at the front door.

5 Many dreams belong to one of the following common types.

– flying   – taking a test
– meeting someone important
– missing a train or plane
– giving a speech in public

Have you ever had one of these types of dream? Can you remember your strangest dream? Talk or write about it, using the past simple and past continuous tenses and some of the words or expressions in 3.

32 | *Time travellers*

Past continuous (2) : *while* and *when*

VOCABULARY AND READING

1 You're going to read a story called *Time travellers*. Here are some of the words from the story. Check you understand what they mean. What do you expect the story to be about?

sightseeing grounds palace crowd favourite atmosphere strange shiver discover eighteenth century lose one's way cottage sculpture uneasy path evil angry disappear bridge lawn grass anniversary prison invade messenger warn guard

2 Read the story and find out why it's called *Time travellers*.

3 Work in pairs. What do you think happened to the two women?

4 Read the story again and decide where these sentences go. The sentences are in the order in which they appear in the story.

a While they were visiting the Palace, one of the women had an idea.

b While they were walking through the Palace grounds, the atmosphere suddenly changed.

c While they were wondering which way to go, they saw a wooden hut under some trees.

d While they were walking past, she stared at them in a very royal way.

e While she was relaxing at the Petit Trianon, an angry crowd of people from Paris invaded the Palace.

GRAMMAR

Past continuous (2): *while* and *when*

You can use *while* **+ past continuous to talk about something that was in progress at the time something else happened or interrupted it. You need a comma at the end of the** *while* **clause.**

While they were visiting the Palace, one of the women had an idea.

You can also put the *while* **clause at the end of the sentence. You don't need a comma.**

One of the women had an idea while they were visiting the Palace.

You can also use *when* **in the past simple clause.**

They were visiting the Palace when one of the women had an idea.

You use *when* **+ past simple to describe two things which happened one after the other. The second verb is often in the past simple.**

When the women got closer, they saw some people in eighteenth-century clothes.

You can also say:

The women saw some people in eighteenth-century clothes when they got closer.

1 Write full answers to these questions.

1 Who did they notice while they were walking to the Petit Trianon?

2 What happened while the evil-looking man was staring at the two women?

3 Who did they see while they were crossing the lawn in front of the Petit Trianon.

2 Rewrite your answers in 1 using *when* + past simple.

3 Write full answers to these questions.

1 What did the two men do when they asked them the way?

2 How did Miss Jourdain feel when she saw the woman and the young girl?

3 What did they see when they crossed the bridge?

4 What did they do when the tour was over?

SPEAKING

1 Work in pairs. Would you like to travel in time? What time would you like to travel to? Talk about the people you'd like to meet and things you'd like to do.

2 Talk about your time travelling with the rest of your class. Find someone who'd like to travel with you. Decide who has got the strangest reasons for travelling.

Time travellers

One hot August day in 1901, two young English women were sightseeing at the Palace of Versailles near Paris. It was extremely hot, and there were crowds of people. 'Why don't we walk to the Petit Trianon?' suggested Miss Anne Moberly. Her friend, Miss Eleanor Jourdain agreed. The Petit Trianon was one of Queen Marie-Antoinette's favourite places. Suddenly everything felt strange. The sun was still shining but Miss Moberly shivered. In the distance they noticed some people. But when the women got closer they saw the people were wearing clothes from the eighteenth century.

They went up to two men wearing long green coats and red three-cornered hats, who were talking to each other. 'Excuse me, gentlemen, but we have lost our way. Could you tell us how to get to the Petit Trianon?' The men turned and looked at the two women in surprise. One of them said, 'Go straight down this path, madam,' and carried on talking.

Down the path was a small cottage. While they were passing it, they looked in through an open door. Inside there was a woman and a young girl. Both were wearing eighteenth-century clothes and were standing very still, like sculptures. Miss Jourdain felt uneasy because she knew she wasn't dreaming.

They came to a point where the path branched left and right. An evil-looking man was sitting on the grass in front, staring angrily at the two women. Suddenly a second man ran up to them and called out to them, 'Don't go that way. Return to the house.' He looked over his shoulder and disappeared again.

The two women took the path the man suggested and crossed a little bridge. Beyond the trees, they could see the Petit Trianon. While they were crossing the lawn in front of the house, they saw a beautiful woman wearing a summer dress, sitting alone on the grass. They went up the steps onto the terrace and joined a group of tourists who were visiting the building. When the tour was over, they returned to Versailles for tea.

While they were walking round the palace, neither lady talked about the strange event. Then a week later, Miss Jourdain suddenly said, 'Do you think the lady in the summer dress was Marie-Antoinette?' The day they were there was 10 August, the anniversary of the day in 1789 when Marie-Antoinette went to prison. Perhaps the man running to tell them to return to the house was a messenger who was coming to warn the Queen about the danger. The two men in green coats and red hats were probably members of the Swiss guard who looked after the Queen while she was staying at the Petit Trianon, and the woman and young girl were possibly the wife and daughter of one of the royal gardeners. And was the evil-looking man the Count de Vaudreuil, who told the crowd where Marie-Antoinette was?

On a second visit to Versailles, Miss Jourdain discovered that the wooden hut was no longer there, nor was the bridge, although old maps showed they were once there.

So what really happened on that hot August day in 1901? Both Miss Jourdain and Miss Moberly believed that, somehow, they travelled back to the Court of Versailles in August 1789. Or was it just a dream caused by the heat of the day?

33 | *Is there a future for us?*

Expressions of quantity (2): *too much/many, not enough, fewer, less* and *more*

READING

1 Read *Is there a future for us?* and underline anything you think is a good suggestion and correct anything you think is wrong.

2 Match the words from the passage in list A with their meanings in list B.

A fumes
muck up
squashed
tandem
greedy

B wanting more than it needs
damage
flat
unpleasant smoke or gas
bike with two seats

3 Which of these statements about the environment do you think Henrietta and Eryn agree with?

1 There's too much noise.
2 The air in the cities isn't clean enough.
3 There isn't enough farmland.
4 There aren't enough clean rivers.
5 We should use less fuel.
6 There are too many people.
7 The sea is too polluted.
8 We should give more money to poor people.
9 We should cut down fewer trees.
10 We should build more houses.

4 Put a tick (✓) by the statements you agree with.

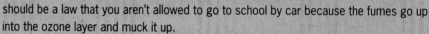

Is there a future for us?

Henrietta (aged 8): The biggest problem with the environment is the ozone layer; there's a hole, and it's getting bigger. It's made by cars and aeroplanes – things which give off fumes. There should be a law that you aren't allowed to go to school by car because the fumes go up into the ozone layer and muck it up.

Eryn (also aged 8): The ozone layer's like a piece of paper covering a rock. It's meant to protect us. I'm scared the hole will get bigger and move around the world and people will get cancer. I think about it sometimes in bed. That could happen in our time. We should stop using aerosols, because they get squashed and all the poisons come out. And we could use horses instead of cars.

Henrietta: We could get tandems, and longer bikes, so children could ride on the back. Cars should be very, very expensive. I heard a woman on the radio saying she didn't care about the ozone layer because she wouldn't be alive, but it's our family who will die.

Eryn: You also get bad pollution from burning down the rainforest. People should leave things alone if they don't own them. There should be fines for destroying the rainforest. And we should give money to poor people in Africa and places.

Henrietta: We should spread out the people evenly. We could say, 'Put your hands up all those who want to live in Africa.' And then we could spread out the food. There's enough to go round. We could easily grow it. England's a very rich and greedy country.

Eryn: We use up far more of the earth than people in Africa, so it's a good idea for the whole world to discuss the environment, but not prime ministers, because they always vote for their own side. The Queen should go instead.

Adapted from *The Independent on Sunday*

GRAMMAR

> **Expressions of quantity (2)**
>
> ***Too much/many* + noun**
> *There's **too much** noise.*
> *There are **too many** people.*
>
> ***Too* + adjective**
> *The sea is **too polluted**.*
>
> ***Not enough* + noun**
> *There isn't **enough** farmland.*
> *There aren't **enough** clean rivers.*
>
> ***Not* + adjective + *enough***
> *The air isn't **clean enough**.*
>
> ***Fewer, less* and *more***
> **You use *fewer* and *more* with countable nouns.**
> *In Britain there are **fewer** men than women. There are **more** women than men.*
>
> **You use *less* and *more* with uncountable nouns.**
> *There's **more** pollution these days. There's **less** clean air.*

1 Complete these sentences with *too, much, many* and *enough*.

  1 It's ___ quiet for me in the countryside.
  2 There aren't ___ forests in the world.
  3 The centre of my country is ___ flat. I prefer the mountains.
  4 There are too ___ factories.
  5 It isn't peaceful ___ in the city.
  6 There's ___ much pollution in the sea.

2 Give your opinion about these issues. Write sentences with *fewer, less and more* beginning *We need...*

cars industry clean water pollution factories farmland fields forests
We need fewer cars.

3 Work in pairs. Find out if your partner agrees with the statements in *Reading* activity 3.

4 Write sentences saying what you don't like about the environment near your home.
There's too much noise.

VOCABULARY AND SPEAKING

1 Work in pairs and look at the words in the box below. Which adjectives and nouns go together?

> beach city cliff coast countryside desert factory farmland field flat forest
> hill hilly industrial industry island jungle lake mountain mountainous
> noisy ocean peaceful poor quiet region rich river rural sea town village

a quiet beach, a flat field

Which words in the box can you use to talk about your country? Can you think of other words to describe your country?

2 Describe different regions of your country using the words in 1 and the words below.

There's a desert in the centre of the country. It's quite rural in the north.

3 Work in pairs. In your opinion, where is the ideal place to live in your country?
I think it's probably somewhere on the West coast, not far from the mountains.

4 Work in groups of three or four and talk about what will happen to our environment in the future. Think about the following:

– the population
– the sea
– the sea level
– the countryside
– towns
– the climate
– the air

Choose from these phrases:

grow bigger grow smaller
become dirtier become cleaner
warm up cool down
improve deteriorate rise fall
get more crowded
get more deserted

The population will grow bigger.

5 Find five predictions about the environment which most people in the class agree with.

The Day of the Dead

It's the end of October, and Mexico is preparing to celebrate the Day of the Dead on 1 November. In cities throughout the country for several weeks before the festival begins, street markets and shops are filled with symbols of death.

It's the highlight of the year of the year for all Mexicans: the day when dead spirits return to the land of the living. But there's nothing sad about this festival. It simply reflects the Indian belief that death is a natural part of life.

November 1 is known as the feast of All Saints and All Souls. The Day of the Dead is called *todos santos* or *dia de muertos* by the Mexicans and the festival usually involves two days of celebration on 31 October and 1 November. Mexicans believe that on the Day of the Dead the souls of dead relatives will return. The Indian festival became a Catholic one when the Spanish brought their religion to Mexico.

A feast is prepared for the dead with their favourite food and drink, cigarettes, sweets and fruit. A special kind of bread, known as *pan de muertos* ('bread of the dead') is baked, traditionally by the men – either the head of the family or the closest relative of the dead person. Today, however, the *pan de muertos* is often bought in markets. A bowl of water and a cloth is put on the table so that the spirits can wash their hands, and sometimes a favourite possession of the dead person is left.

The dead person is not usually seen when they return, but their spirit is felt by the family. After the festival, the food is given to the community, and the gifts are arranged around a wooden frame which is decorated with coloured papers, flowers and fruit.

The island of Janitzio is famous for its Day of the Dead celebrations, and has become a major tourist attraction. Just before midnight on 1 November, the lake which surrounds Janitzio is lit up by hundreds of torches. These show the route of the *lancha*s (small boats) which carry the families to the island. They go with their gifts to the cemetery where they will spend the night.

The cemetery is crowded not only with family and friends, but with tourists, photographers and even film crews. But later in the night, the tourists leave, and the families remain until morning. Through the night, attracted by the light of the candles and perfume of incense and flowers, the souls of the dead return once more to their families.

Adapted from *BBC World magazine*

READING AND VOCABULARY

1 You are going to read about Mexico's Day of the Dead celebrations. Which of the words in the box do you expect to see?

> belief Buddhist candle Catholic celebration
> cemetery dead highlight mosque Muslim perfume
> saint soul symbol temple torch tourist

2 Read *The Day of the Dead* and decide what is the most surprising or interesting piece of information in the passage.

3 Work in pairs. Read the passage again. Decide what the photos show and write a caption for each one.

GRAMMAR

> **Present simple passive**
> **You form the present simple passive with *am/is/are* + past participle.**
> *A splendid feast **is prepared**.*
> *Shops **are filled** with symbols of death.*
>
> **Questions**
> *What **is** November 1 **called**?*
>
> **You use the passive when you are more interested in the object of the sentence or you don't know who or what does something.**
> *A bowl of water is put on the table.*
> **(You aren't interested in who put it there.)**
> *Shops are filled with symbols of death.*
> **(You don't know who filled the shops.)**
> **If you are more interested in the object, but you know who or what does something, you use *by*.**
> *A special kind of bread is baked **by** the men.*
> **If you are more interested in the subject, you use an active sentence.**
> *The men **bake** a special kind of bread.*

1 Here are some answers to questions about the passage. Write the questions using the passive.

1 The Day of the Dead.	5 The dead person's spirit.
2 A feast.	6 The food.
3 A special kind of bread.	7 The gifts for the dead.
4 A bowl of water and a cloth.	8 The lake surrounding Janitzio.

What is November 1 called ?

2 Rewrite these sentences in the passive.

1 They celebrate All Souls' Day on November 1.
2 The Mexicans call it the Day of the Dead.
3 They prepare a feast.
4 They often buy the bread in markets.
5 The family feels the spirit.
6 They arrange the gifts around a wooden frame.
7 Boats carry the families to the island.
8 The lights of the candles attract the souls of the dead.

All Souls' Day is celebrated on November 1.

3 Look at these pairs of active and passive sentences. In each pair, which do you think is the better sentence?

1 a They make Fiat cars in Italy.
   b Fiat cars are made in Italy.
2 a They sell Macdonalds hamburgers in many countries.
   b Macdonalds hamburgers are sold all over the world.
3 a My mother does a lot of cooking.
   b A lot of cooking is done by my mother.
4 a My brother writes poetry.
   b Poetry is written by my brother.
5 a Coffee is drunk in most countries.
   b They drink coffee in most countries.
6 a Cats like warm beds.
   b Warm beds are liked by cats.

SPEAKING AND WRITING

1 Think about an important ritual or festival in your country. Make notes on:

– what it's called	– what is celebrated
– what preparations are made	– what food is cooked
– what is drunk	– what gifts are given

2 Find someone in your class who has chosen a different ritual or festival. Ask questions about it and make notes, and answer questions about the one you chose.

What is it called?   *The Palio.*
When does it take place?   *On two days in July and August.*

3 Write a passage describing the ritual or festival your partner chose.

The highlight of the year for people from Siena is called the Palio. It takes place on...

Making comparisons (3): *but, although, however*

VOCABULARY AND READING

1 Look at the words in the vocabulary box. Group the words under these headings: *things to eat, things on the table, things to cook with, parts of the body* and *things to say when eating or drinking.*

> bowl cheers chips chin cup
> dish elbow fork hand
> ice cream jam knife lap
> melon napkin neck pasta
> plate pot saucepan saucer
> sausage spoon steak
> table cloth teaspoon toast

things to eat: chips...

2 Work in pairs and check your answers. Think of two or three other words which you can add to each group. (You can use a dictionary if necessary.)

3 Read the questions in *Mind your manners!* and think about your answers.

4 Work in pairs and discuss your answers.
In my family we have dinner at ten o'clock.

5 Tell the rest of the class what your answers are. Use these expressions.

Both of us think that...
Neither of us thinks that...
I think ... but Elena thinks...

Mind your manners!

1 What do you say at the start of a meal?

2 What time do you have lunch and dinner?

3 How long does a typical lunch or dinner last?

4 Do you usually use a knife and fork? If so, which hands do you hold them in?

5 Do you use a napkin? If so, where do you put it?

6 At which meals do you eat the following food?
melon pasta fish steak

7 Where do you put your knife and fork when you have finished your meal?

8 Where do you put your hands when you're at the table but not eating?

9 Do you eat cake with a fork or a spoon?

10 What food do you often eat with your fingers at the dining table?

11 When do you usually drink coffee and tea?

12 When can you smoke during a meal?

13 What do you say and do when someone raises their glass?

14 Do you have soup in the summer?

15 Do you eat salad in the winter?

LISTENING AND SPEAKING

1 Work in groups of three. You're going to hear Stephen, who is English, talking about table manners.

Student A: Turn to Communication activity 6 on page 99.
Student B: Turn to Communication activity 4 on page 98.
Student C: Turn to Communication activity 14 on page 100.

2 Now work together and talk about Stephen's answers to the questions in *Mind your manners!*

[cassette icon] Listen again and check.

GRAMMAR

Making comparisons (3): *but, however, although*
You use *but, however* and *although* to make a comparison which focuses on a difference. You put *but* at the beginning of a sentence or to join two sentences.
*We drink coffee in the morning. **But** we don't drink it in the afternoon.*
*We drink coffee in the morning **but** we don't drink it in the afternoon.*

You use *although* at the beginning of a subordinate clause. You need to separate the subordinate and the main clause with a comma.
*We usually have dinner at six, **although** some people have dinner later.*

You can put the subordinate clause at the beginning or at the end of the sentence.
***Although** we usually have dinner at six, some people have dinner later.*

You use *however* at the beginning of a sentence. It is followed by a comma.
*We drink coffee in the morning. **However,** we don't drink it in the afternoon.*

1 Complete these sentences with *however* or *although*.

1 Many people have dinner quite early, ___ we eat quite late, at about nine.
2 Most people have milk in their tea, ___ I prefer lemon.
3 I hold my fork in my left hand to cut food, ___ I change to my right hand to eat.
4 I usually drink coffee in the morning, ___ I sometimes have a cup after dinner.
5 We eat melon at the start of a meal. ___ , some people have it at the end.
6 You don't usually smoke while you're eating. ___ , it's OK to smoke after the meal.

2 Write sentences making comparisons between table customs in your family and Stephen's.

SOUNDS

[cassette icon] Listen to the sentences in *Grammar* activity 1. Notice the stress and intonation of sentences with *however* and *although*.

Now say the sentences aloud.

SPEAKING AND WRITING

1 Work in groups of three and write down three examples of table manners – two true and one false.
You should always eat thick soup with a knife and fork.
In a restaurant, you can attract the waiter's attention by whistling.
After a meal, say goodbye and shake hands with all the other people.

2 Present your three examples of table manners to the rest of the class. Try to add as much information and context as possible.

The rest of the class must guess which is the false one.

Progress check 31-35

VOCABULARY

1 Some verbs are followed by a particle (an adverb or a preposition). These are called multi-part verbs.

take out   She took out a pen.
listen to   He's listening to the radio programme.

Use your dictionary to find out how many multi-part verbs you can make with the verbs and the adverbs or prepositions below.

Verbs   come   fall   look   pick   stand   take   turn   throw

Particles   at   away   down   in   off   round   up

2 Now complete these sentences with a suitable particle from the list above.

1 He took ___ his coat.
2 I picked ___ my wallet.
3 She fell ___ the stairs.
4 I threw the wrapping ___ .
5 He turned ___ and looked ___ me.
6 She stood ___ when we came ___ .

3 With some multi-part verbs, you can put the noun object after or before the particle.

take out   She took out a pen. She took a pen out.
put on   He put on his hat. He put his hat on.

But you must put the pronoun object between the verb and the particle.

She took it out.   He put it on.

Rewrite these sentences, replacing the noun with a pronoun.

1 Don't throw the newspaper away.
2 He turned the radio down.
3 He took his hat off.
4 He washed the dishes up.
5 He gave up smoking.
6 I'll find out her name.

4 With other multi-part verbs you always put the object after the particle.

listen to   She's listening to the radio programme. She's listening to it.
look at   He looked at the old woman. He looked at her.

Rewrite these sentences with pronouns.

1 She thought about John.
2 We voted for the socialists.
3 I waited for my mother.
4 He looked at the picture.
5 She listened to the radio.
6 We looked after their cat.

5 Look at the vocabulary boxes in Lessons 31 – 35 again. Choose words which are useful to you and write them in your *Wordbank*.

GRAMMAR

1 Join the two parts of the sentence with *when*.

1 I/play tennis/hurt my leg.
2 He/walk to work/see his friend.
3 They/watch television/fall asleep.
4 She/talk to me/start to cry.
5 We/sit in garden/hear a loud noise.
6 I/look for a pen/find some money.

1 I was playing tennis when I hurt my leg.

2 Rewrite the sentences in 1 using *while*.

1 While I was playing tennis I hurt my leg.

3 Write questions.

1 What/do/8am yesterday?
2 Who/talk to/last night?
3 Why/work so hard/last week?
4 What/talk about/this morning?
5 Where/have dinner/last Sunday?
6 Why/laugh/just now?

1 What were you doing at 8am yesterday?

4 Rewrite the sentences with *too* + adjective or *not* + (adjective) *enough.*

1 It's too noisy in here. (quiet)
2 It wasn't dark enough to sleep. (light)
3 I was too cold. (warm)
4 The region is too industrial. (rural)
5 The hotel room was too dirty. (clean)
6 The jacket wasn't big enough. (small)

1 It isn't quiet enough in here.

5 Write correct answers to these sentences.

1 Is champagne made in Britain?
2 Are shops closed during the weekend?
3 Is leather used for making clothes?
4 Is tea drunk at night?
5 Are grapes eaten at breakfast?
6 Is the housework done by men?

1 No, it's made in France.

6 Rewrite these sentences in the passive. Use *by* if necessary.

1 They drink a lot of tea in Britain.
2 They eat a lot of meat in Argentina.
3 The Spanish eat dinner at 10pm.
4 You buy medicine at the chemist's.
5 The dentist examines your teeth.
6 You use newspapers for wrapping things.

1 A lot of tea is drunk in Britain.

7 Join these sentences about Catherine, an English woman, by rewriting them with *although*.

1 She likes coffee. She prefers tea.
2 She doesn't usually have time to eat in the mornings. At weekends she has a large breakfast.
3 She usually has a sandwich for lunch. She sometimes has a salad when she goes out with friends.
4 She doesn't smoke. She doesn't mind other people smoking in her home.

8 Rewrite the sentences in 7 with *however*.

SOUNDS

1 Complete these words with *v* or *w*.

_ ait _ allet mo _ e _ isit _ ant _ ork sand _ ich
dri _ e in _ ite _ omen _ ear shi _ er _ alk beha _ e

Listen and check. Say the words aloud.

2 Listen and tick (✓) the word you hear.

1 ear hear   2 air hair   3 eye high   4 at hat
5 eight hate 6 eat heat   7 art heart 8 as has

Now say the words aloud.

3 Listen to the answers to *Vocabulary* activity 3. Notice how the speaker stresses the verb and the particle.

1 Don't throw it away.   4 He washed them up.
2 He turned it down.   5 He gave it up.
3 He took it off.   6 I'll find it out.

Now say the sentences aloud.

WRITING

1 Look at some rules for punctuation in English.

You use commas:
– to separate main and subordinate clauses.
While they were waiting, it started to rain.

– to separate signpost words and phrases, such as *however, as a result, in fact, first, next, after that* from the rest of the sentence.
I like coffee. However, I prefer tea.

Rewrite this passage with the correct punctuation. (You may like to look at *Progress check lessons 1 – 5, Reading and writing* before you begin.)

an english couple drove to paris for a holiday they found their way to the hotel but couldnt find anywhere to park finally they found a space but by now it was dark while they were trying to park a parisian saw their british car and offered to help you have a meter behind you he shouted so the englishman moved back a little the man shouted again you have a meter behind you the driver moved a little more the frenchman waved and shouted you have a meter behind you the englishman who was getting a little angry by now reversed the car rather quickly there was a crash the parisian pointed at a post which was lying under the car i said you had a meter behind you

2 Read the story again and decide where these words can go. Some words can go in more than one position.

short back parking friendly loud metal

3 Think of six more words which could go into the story and write them down. Now work in pairs. Show your words to your partner and ask him/her to decide where they can go in the story.

36 | *Lovely weather*

Might and *may* for possibility

READING

1 Read *When's the best time to visit your country?* and make a list of the best times to visit each country.

2 Work in pairs. Is your country mentioned? Do you agree on the best time to visit? If it isn't mentioned, say when the best time to visit is and add it to your list.

VOCABULARY

1 Look at the box and underline the words for seasons. Which of the other words in the box go with each season in your country?

> autumn dry hurricane flood
> sun freezing storm humid
> rain changeable ice lightning
> wind mist spring snow frost
> summer fog thunder wet
> mild winter

winter: freezing, fog, ice

Which words don't you use to talk about your weather?

2 Which of the nouns in the vocabulary box can you turn into adjectives?
fog – foggy

When's the best time to visit your country?

'May or October is best because it's not too humid, although there may be a few showers.' *Wang Wei, Hong Kong*

'We live in New England, so the best time is late August and September when the weather starts to get cooler and the leaves on the trees change colour.' *Norman, USA*

'July and August are fabulous. It might be quite cool but it'll be very pleasant. But don't come in the winter, it'll be dark all day and freezing cold.' *Ingrid, Sweden*

'We say the spring is the best time to visit us. But it may rain and there's quite a lot of wind, so you may miss the cherry blossom. Come in the autumn, the autumn leaves will be beautiful.' *Reiko, Japan*

'Come to Prague in winter and you'll see how beautiful the city is. There may be snow and there are fewer tourists.' *Frantisek, Czech Republic*

'Have you ever had Christmas dinner on the beach? Come in December, and you won't see a cloud in the sky.' *Kevin, Australia*

'In the summer, it's very hot and humid on the coast, although you might find it a little cooler in Mexico City because it's high up.' *Miguel, Mexico*

'Spring and autumn are the best, the climate is perfect on the Aegean and Mediterranean coasts, although it may be cooler in central Anatolia. You won't get much rain between May and October.' *Yildiz, Turkey*

'In December you'll find it a bit like England in July, not too hot. And it'll be rainy and cold in the winter. But it depends where you want to go. It's a big country and very mountainous.' *Maria Sara, Chile*

GRAMMAR

> ### Might and may for possibility
> **Might** and **may** are modal verbs. Remember that they have the same form for all persons and you don't use the auxiliary *do* in questions and negatives. You can use **might** or **might not** + infinitive to talk about possible future events.
> *It **might** rain tomorrow.* *There **might not** be much sun.*
>
> You can also use **may** or **may not**.
> **May** is a little more sure than **might**.
> *It **may** rain tomorrow.* *There **may not** be much sun.*
>
> Remember you use **will** or **won't** to make predictions.
> *It'll be dry and sunny tomorrow over the whole country. There **won't** be much wind.*

1 Imagine you're talking to a friend who wants to visit your country. Give the following advice and explain why. Use *because* + *might*.

1 bring an umbrella	4 get some sun-tan lotion
2 wear warm clothes	5 buy a good map
3 bring your camera	6 bring a swimsuit

You should bring an umbrella because it might rain.

2 Choose the correct form of the verb.

1 *I'm going/I might go* to Kenya next week. I've got my ticket.
2 *I've booked/I may book* into a hotel. I've got a room at the Ambassador.
3 *I'm going /I might go* to Lake Victoria or perhaps to the beach near Mombasa.
4 *I won't/I might not* spend more than a week on the coast, because I want to see the National Park.
5 *I'm going/I might go* on safari if there's room for me.
6 *I'm coming/I might come* home on the sixteenth. It's the day before I get married.

3 Say what the weather might or might not be like:

– tomorrow	– next week
– next month	– in ten years' time

4 Think about next weekend. Say what you might or might not do if...

1 ...it's raining	4 ...it's sunny
2 ...it's snowing	5 ...you have work to do
3 ...it's cold	6 ...you feel tired

1 I might stay at home and watch TV.

WRITING

1 Delphine wrote to her friend Benita asking for some advice. Read Delphine's letter and Benita's reply, and find the answers to these questions.

a What type of accommodation might Delphine like?
b What type of clothes should Delphine bring and why?
c When might be the best time to come?
d What might Delphine like to do?
e What might the weather be like?

> Dear Benita,
> At last I'm thinking of coming to Britain. When's the best time to come? Thanks for your help!
> Love, Delphine

> Dear Delphine,
>
> It's good to hear you may want to visit Britain. I think October is the best time to come. It's usually quite warm, although there may be some rain, and it might be a bit cold in the evening, so bring a sweater. Best of all, there won't be so many tourists. Make sure you bring an umbrella. You may like to stay in a bed and breakfast, which is a private house with a guest bedroom. You won't have to pay much for this type of accommodation. When you arrive in London you may like to buy a railcard to travel around the rest of the country.
>
> I hope this is useful.
>
> Best Wishes, Benita

2 Number the questions in the order that Benita answers them.

3 Imagine an English friend wants to visit your country and has asked you when the best time is to visit. Make notes answering the questions in 1.
Accommodation: stay in a bed and breakfast...

4 Write a letter to an English friend who wants to visit your country. Use the letter in 1 to help you and your notes. Make sure you answer the questions in full sentences, and in the order you numbered them in 2.

37 *Help!*

First conditional

VOCABULARY AND SPEAKING

1 Look at the words in the box. Which can you use to describe the incident in the picture?

accident electricity steal break victim kill button
catch fire consulate dangerous passport shock
ambulance wallet fire flood injured package gas
ground floor gun burn mugger plug plug in press
rescue drown driving licence bomb switch on
switch off explode witness unplug burglar

2 Put the words in the box with the following incidents. Many can go with more than one situation.

a an emergency at home
b a road accident
c a bomb alert
d a mugging
e a flood

Check your answers with another student.

3 Look at these sentences. Decide in which incident you might hear them. Can you guess what the underlined pronouns refer to?

1 'If you touch <u>it</u>, you'll get a shock.'
2 'If you don't give <u>it</u> back, I'll call the police!'
3 'If <u>it</u> doesn't stop rising, we won't be able to escape.'
4 'If you go to the consulate, they'll give you another <u>one</u>.'
5 'If you lose <u>them</u>, <u>they</u>'ll replace them in a couple of days.'
6 'If you light a match, <u>it</u>'ll cause an explosion.'
7 'If we call an ambulance, <u>they</u>'ll take her to hospital.'
8 'If no one picks <u>it</u> up, I'll call the guard.'

GRAMMAR

> **First conditional**
> You use the first conditional to talk about a likely situation and to describe its result. You talk about the likely situation with *if* + present simple. You describe the result with *will* or *won't*.
> You separate the two clauses with a comma.
> *If you **give** it to me, I'**ll let** you go.*
> *If you **don't give** it back, I'**ll call** the police.*
> *If it **doesn't stop** rising, we **won't be able to** escape.*
> You often use the first conditional for promises, threats or warnings.

1 Match the two parts of the sentences.

1 If you unplug the machine,
2 If you give me the money,
3 If we don't escape,
4 If they give you another passport,
5 If they replace your traveller's cheques,
6 If there's an explosion,
7 If she goes to hospital,
8 If the guard thinks it's a bomb,

a ...there won't be any trouble when you leave the country.
b ...you'll be able to carry on with your holiday.
c ...I'll let you go.
d ...we'll drown.
e ...you won't get a shock.
f ...they'll mend her broken leg.
g ...he'll ask everyone to leave the train.
h ...there'll be a lot of damage.

2 Complete the sentences.

1 If I have enough money next summer,...
2 If I have time this week,...
3 If my friends are doing nothing tonight,...
4 If I work hard next week,...
5 If I finish this book,...
6 If I learn English,...

SOUNDS

1  Listen and repeat these phrases.

I'll   I'll do that   she'll   she'll show you
you'll   you'll like it   we'll   we'll love it
he'll   he'll call you   they'll   they'll learn it

2 Listen and mark the words the speaker links.

1 If you touch that, it'll explode.
2 If no one picks it up, I'll do it.
3 If you don't stop eating, you'll be ill.
4 If you don't let go, I'll scream.
5 If you don't give it to me, I'll tell your uncle.
6 If you ask her, she'll answer.

Now say the sentences aloud.

LISTENING AND SPEAKING

1 You're going to hear Kate, an Australian woman, describing an incident in three parts. Listen to the first part and decide what the incident is. Choose from the list in *Vocabulary and speaking* activity 2.

2 Work in pairs and check you understood what happened.

3 Work in pairs. Look at these extracts from the second part and try and guess what happens next. (The extracts are in the wrong order.)

...they had got some good news ☐
...now it was me who was feeling sorry ☐
...the young man wasn't Australian ☐
...admitted he was guilty ☐
...the bank clerk called the police ☐
...was unemployed and had a family to look after ☐
...a young man was trying to change some Australian money ☐
...'I'm sorry, I'm really sorry.' ☐

Now listen to the second part of the story and number the extracts as you hear them.

4 Work in pairs. Predict the possible results of these situations. There may be more than one result.

1 If the police let the man go,...
2 If the man gets a fine,...
3 If the magistrate sends the man to prison,...

5 Work in pairs. How do you think the story finishes?

Listen to the third part and find out. Do you think Kate made the right decision?

6 Have you ever been involved in an incident like Kate? Tell your partner about it.

38 *My perfect weekend*

***Would* for imaginary situations**

We asked Stephen from Leeds, and Paula from Nottingham about their perfect weekend. Here are the questions and their answers.

1 Where would you go?

2 How would you travel?

3 Where would you stay?

4 Would you take a companion?

5 What essential piece of clothing or kit would you take?

6 What would you have to eat and drink?

7 Would you take anything to read?

8 Who would be your least welcome guest?

9 What luxury would you take?

10 What three things would you most like to do?

Stephen

a My lap-top computer.

b My swimming trunks, because I love the sea.

c Well, I'd like to take my dog, but I wouldn't be allowed to bring him back to England. So I'd go on my own.

d Sit on the beach, have a good lunch and read a good book.

e To a small town I know in France, by the sea. But I won't say where!

f Yes, 'The Kingdom by the Sea', by Paul Theroux.

g In a friend's house right by the beach. It's very quiet, a great place to relax.

h The local wine, bread, cheese and those enormous tomatoes you can get there.

i The Prime Minister. He would only make me angry.

j By train. I wouldn't go by plane – I hate it.

Paula

a Good shoes. I expect to be on my feet most of the day.

b At the Waldorf Hotel. It's quite an old-fashioned place, but the service is very good.

c Go shopping, have something to eat ... and then go shopping again.

d Probably not. Maybe I'd pick up a book at the airport, something by Jackie Collins.

e Pastrami on rye, bagels, in fact anything from those amazing delicatessens they have in New York. Oh, and champagne, of course.

f New York. I love the place. The shops are just wonderful.

g My ex-husband.

h By Concorde. More time for shopping!

i Yes, my best friend Shirley. We do everything together. She's the only person who likes shopping more than I do.

j My credit cards. But they're more of a necessity, actually.

VOCABULARY AND SPEAKING

1 Look at the words in the box. Are the things *luxuries* or *necessities*?

> cassette recorder cat central heating champagne comb
> companion computer credit card dog insurance jewellery
> novel orange juice pen religion rice salt seafood soap
> swimming trunks telephone washing machine water

2 Work in pairs and discuss your answers to 1. Now talk about other things that are necessities and luxuries for you.

READING

1 Read *My perfect weekend* and match the questions with Stephen and Paula's answers.

2 Think about your answers to the questions. Would you do the same things as Stephen and Paula?

GRAMMAR

> *Would* for imaginary situations
> **Would** is a modal verb. Remember that modal verbs have the same form for all persons and you don't use the auxiliary *do* in questions and negatives.
> You use *would* to talk about the consequence of an imaginary situation. You often use the contracted form *'d* or *wouldn't*.
>
> Where **would** you go? *I'd go to a small town in France.*
> How **would** you travel? *I **wouldn't** go by plane.*
>
> | **Questions** | **Short answers** |
> |---|---|
> | **Would** you take a companion? | *Yes, I **would**.* |
> | **Would** you take anything to read? | *No, I **wouldn't**.* |

1 Here are some more answers from *My perfect weekend*. Write the questions.

1 'I'd play tennis and maybe go for a swim.'
2 '*Gone with the Wind*. I've seen it five times already.'
3 'Verdi's *Requiem*. It makes me cry.'
4 'Yes, I'd have dinner in a small restaurant in the countryside.'
5 'No, I don't like to read a newspaper when I'm on holiday.'
6 'Sunday night, or if possible, Monday morning.'

2 Write your answers to the questions in *My perfect weekend*. Write full sentences.

SOUNDS

[cassette icon] Listen to these sentences. Notice how you don't always hear the /d/ in *'d* before a verb beginning with /t/ or /d/.

1 I'd take my friend. 4 She'd talk to her friends.
2 He'd do some work. 5 We'd drink champagne.
3 I'd tell you a story. 6 They'd travel by plane.

Now say the sentences aloud.

LISTENING

1 [cassette icon] Listen to six people answering a question about their perfect weekend. Put the number of the question they answer by the name of the speaker.

Alex	☐	Barbara	☐	Alan	☐
Daniel	☐	Emma	☐	Jane	☐

2 Work in pairs. Say what their answers were.
Alex would drink coconut milk.

[cassette icon] Listen again and check.

SPEAKING

1 Work in pairs. Ask and answer the questions about your perfect weekend. Ask for extra information about each question, using *would*.
Where would you go?
I'd go to my grandmother's house.
Why would you like to go there?
Because she would be pleased to see me.

2 Choose two or three of the people in your class and imagine how they would answer the questions.
I think Marco might go to the beach.
Yes, and he might go by motorbike.

3 Now ask the people you chose in 2 and find out if you guessed correctly.

39 | *The umbrella man*

Second conditional

GRAMMAR

> ### Second conditional
>
> **You use the second conditional to talk about an imaginary or unlikely situation and to describe its result. You talk about the imaginary or unlikely situation with *if* + past simple. You describe the result with *would* or *wouldn't*.**
>
> *If I **had** a lot of money, I **would** give some away.*
> *If a stranger **asked** me for money, I **wouldn't** give him any.*
>
> **You form the second conditional with *if* + past tense, *would* + infinitive. You separate the two clauses with a comma.**
>
> *If I found some money in the street, I'd keep it.*

1 Match the two parts of the sentences.

1	If I was extremely rich,	a	I'd go to Jamaica.
2	If I won a free holiday,	b	I'd give up my job.
3	If I lost my wallet,	c	I'd live in New York.
4	If I spoke English fluently,	d	I'd go to the police.
5	If I changed my job,	e	I'd get a better job.
6	If I could live anywhere,	f	I'd be much happier.

2 Write sentences saying what you would do if...

– a stranger asked you for money
– you didn't have any money
– someone lied to you
– someone offered you an expensive gift

READING AND LISTENING

1 Work in pairs. You're going to read a story called *The umbrella man* by Roald Dahl. Here are some words from part 1. What do you think is going to happen?

mother   London   dentist   raining   taxi
passengers   small man   seventy   umbrella   favour
suspicious   help

Now read part 1 and check.

The umbrella man

Part 1 I'm going to tell you about a funny thing that happened to my mother and me yesterday evening. Yesterday afternoon, my mother took me to London to see the dentist. After that, we went to a cafe. When we came out of the cafe it was raining. 'We must get a taxi,' my mother said. Lots of them came by, but they all had passengers inside them.

Just then a man came up to us. He was a small man and he was probably seventy or more. He said to my mother politely, 'Excuse me.' He was under an umbrella which he held high over his head.

'Yes?' my mother said, very cool and distant.

'I wonder if I could ask a small favour of you,' he said. I saw my mother looking at him suspiciously. She is a suspicious person, my mother. The little man was saying, 'I need some help.'

2 Work in pairs. What would you say and do in these circumstances? Would you be suspicious?

3 Work in pairs. Here are some phrases from part 2 of the story. Who do you think is speaking? What do you think is going to happen? Put the phrases in the right order.

a 'Why don't you walk home ?'
b 'He's in some sort of trouble.'
c 'Thank you, madam, thank you.'
d 'I've never forgotten it before,'
e 'I think I'd better just give you the taxi-fare.'
f 'I would never accept money from you like that!'
g 'I'm offering you this umbrella to protect you.'
h 'Are you asking me to give you money?'

[cassette] Now listen to part 2 of the story and check.

4 Work in pairs. Talk about what you would do in the circumstances if you were:

– the narrator   – the narrator's mother   – the old man

What do you think happens next? Read part 3 of the story and find out.

Part 3

'Come under here and keep dry, darling,' my mother said. 'Aren't we lucky! I've never had a silk umbrella before.'
    'Why were you so unpleasant to him?' I asked.
    'I wanted to be sure he was a gentleman. I'm very pleased I was able to help him.'
    'There he goes,' I said. 'Over there. He's crossing the street. He's in a hurry.'
    We watched the little man. When he reached the other side of the street, he turned left, walking very fast.
    'He doesn't look very tired, does he, mummy? He doesn't look as if he's trying to get a taxi, either.'
    My mother was standing very still. 'He's up to something. Come with me.' We crossed the street together. It was raining very hard now, but we were under the silk umbrella.
    'He said he was too tired to walk and now he's almost running.'
    'He's disappeared!' I cried. 'Where's he gone?'

5 Work in pairs and answer these questions.

1 Where has the old man gone?
2 Do you think they will see him again?
3 How do you think the narrator and her mother feel?
4 How do you think the story finishes?

[cassette] Now listen to the last part of the story.

VOCABULARY

Here are some useful words and expressions from the story. Check you know what they mean.

> cafe dentist exchange
> excuse me expect favour
> gentleman happen pleased
> pound put on run silk stand
> still suspicious take out taxi
> trouble umbrella unpleasant

Are there any other words or expressions you would like to add to this list?

WRITING

1 Think about the old man's behaviour. Why do you think he behaved like this? Write a sentence describing what happened from his point of view. Think about what happened before he met the narrator and her mother.
It was a fine day when I went out, but soon it started raining, which was good, because I was feeling rather thirsty.

2 When you are ready, give your sentence to another student and you will receive an opening sentence from someone else. Read it, then write another sentence to continue the story, with as many details as possible. Change stories like this every time you write a sentence, until you finish the story. Write at least six sentences.

How unlucky can you get?

Past perfect: *after, when* and *because*

VOCABULARY AND LISTENING

1 You are going to listen to a story about an unlucky traveller. Here are some words from the story. What do you think is going to happen?

accident	arrest	break down	capital	cost	crash	exhaust pipe	fall off	guard
journey	mend	musician	nightmare	pack	police	queue	rusty	set off	spend

2 Now listen to the story. Did you guess correctly in 1?

3 Work in pairs. Look at these extracts from the story and number them in the correct order.

- ☐ He didn't have very much money left.
- ☐ He decided to go to Britain.
- ☐ He bought his ferry ticket.
- ☐ The police stopped him.
- ☐ He drove for a hundred miles.
- ☐ They let him through.
- ☐ He packed his bags and set off in his car.

- ☐ He had to queue all day at the border.
- ☐ At Calais the car broke down.
- ☐ He drove all night through Hungary.
- ☐ He drove off the boat.
- ☐ In France he crashed the car.
- ☐ The exhaust pipe fell off.
- ☐ The border guards looked carefully at his passport.

4 Listen again and check.

GRAMMAR

Past perfect: *after, when* and *because*

You use the past perfect to talk about one past action that happened before another past action. You often use *after, when* and *because,* and you use the past simple for the second action.

After he **had decided** to go to Britain, he **packed** his bags.
When he **had packed** his bags into his car, he **set off.**
Because he **had spent** so much, he **had** very little money.

You form the past perfect with **had** + past participle.
*The car's exhaust pipe **had fallen** off.*

You often use the contracted form **'d**.
*He'**d** wanted to be a rock musician for years.*

1 Look at these sentences and explain the difference between them.

   a He left the boat when he heard a loud noise.
   b He had left the boat when he heard a loud noise.

2 Choose the best tense.

   1 The people at the border *decided/had decided* to leave their homes.
   2 He felt nervous but the guards *let/had let* him through.
   3 He entered the West and he *drove/had driven* across Southern Europe.
   4 At Calais the car broke down and it *cost/had cost* a lot of money to mend it.
   5 He *spent/had spent* most of his money by the time he got to England.
   6 He heard a loud noise. He saw that the exhaust pipe *fell/had fallen* off.

3 Answer these questions using the past simple.

   1 What did the man do after he'd packed his bags?
   2 What did he do after he'd driven all night through Hungary?
   3 What did he do after he'd entered the West?
   4 What happened after he had left the boat?

4 Answer these questions using the past perfect. Start your answer with the word in brackets.

   1 When did he set off for Liverpool? (After...)
   2 Why did he decide to go to Britain? (Because...)
   3 When did he have to queue all day at the border? (After...)
   4 Why did he have to queue? (Because...)
   5 When did he have a crash ? (After...)
   6 Why did the car make more noise? (Because...)
   7 Why did he stop the car? (Because...)
   8 When did the police see him? (After...)

SOUNDS

📼 Listen and tick (✓) the sentences you hear.

   1 He decided to go to Britain.   He'd decided to go to Britain.
   2 He packed his bags.   He'd packed his bags.
   3 He got to the Austrian border.   He'd got to the Austrian border.
   4 He entered the West.   He'd entered the West.
   5 He arrived in France.   He'd arrived in France.
   6 He had very little money.   He'd had very little money.

Now say the sentences aloud.

SPEAKING AND WRITING

1 Work in pairs. How do you think the story finishes?

2 Rewrite the story joining the events in *Vocabulary and listening* activity 3 with *after, when* or *because* + past participle.
After he'd decided to go to Britain, he packed his bags and set off in his car.

3 Turn to Communication activity 17 on page 100 to find out how the story finishes.

4 Have you ever had an unlucky experience? Make a list of everything that happened, but don't write the events in the order they happened.
   – *missed my bus*
   – *had an argument with my boss*
   – *got to work late*
   – *forgot my briefcase*
   – *went home*

5 Work in pairs. Show each other the list of events. Ask and say the order in which they happened.
Did you miss your bus after you had forgotten your briefcase?
Yes, I did.

6 Take your partner's notes and write a paragraph describing what happened to him/her. Use the past perfect and *after, when* and *because*.

Progress check 36-40

VOCABULARY

1 Look at these phrases with *make* and *do*.

do   do well   do harm   do business   do the washing up
do the housework   do your homework   do the shopping

make   make the bed   make the coffee   make a phone call
make a mistake   make a noise

You can put many words and phrases with *make* and *do*. *Make* often suggests creating something, and *do* often suggests work, but there are many exceptions.

Use your dictionary and find out if these words and phrases go with *make* or *do*.

an appointment   an arrangement   a cake   notes   friends   the cleaning
damage   your best   the ironing   a decision   a cup of tea   someone a favour

2 You use adverbs to describe verbs. You form an adverb by adding *-ly* to the adjective.
peaceful – peacefully   even – evenly   appropriate – appropriately

Adjectives ending in *-y* drop the *-y* and add *-ily*.
noisy – noisily   tidy – tidily

Adjectives ending in *-le* drop the *-e* and add *-y*.
comfortable – comfortably   sensible – sensibly

The adverb of *good* is *well*.
She speaks good Italian. She speaks Italian well.

Fast, hard, late, and *early* are both adjectives and adverbs.

Make adverbs from these adjectives.

cultural   local   slow   quick   heavy   quiet   expensive   beautiful
interesting   formal   fashionable   tidy   healthy   nervous   polite   happy

3 Complete these sentences with an adverb from 2.

1 The rain fell very ___ .
2 She was singing very ___ .
3 He dressed very ___ .
4 He spoke rather ___ .
5 It was raining so we drove very ___ .
6 We live ___ , not far from here.

4 Look at the vocabulary boxes for lessons 36 – 40 again. Choose words which are useful to you and write them in your *Wordbank*.

GRAMMAR

1 Rewrite these sentences with *might (not)*.

1 It will possibly rain today.
2 It is possible that I will stay a week.
3 She possibly won't ring me tonight.
4 It is possible that the plane won't leave on time.
5 He will possibly arrive soon.
6 It will possibly be less noisy.

1 It might rain today.

2 Write six things you may or may not do this weekend.

3 Make sentences with *if*.

1 leave now/catch your train
2 stay in bed/feel better
3 work hard/get a good job
4 eat carrots/be able to see in the dark
5 go shopping/spend a lot of money
6 ride a bike/save energy

1 If you leave now, you'll catch your train.

4 Write questions about your perfect day.

1 Where/go?
2 What/do in the morning?
3 Where/go in the afternoon?
4 How/get there?
5 What/do in the evening?
6 When/go home?

1 Where would you go?

5 Write answers to the questions in 4 about your perfect day.

6 Match the two parts of the sentence.

1 If you took the train,
2 If he left now,
3 If they bought a house in France,
4 If I learned English,
5 If you visited my parents,
6 If we moved house,

a ...they'd be pleased to see you.
b ...he'd be home by seven.
c ...we wouldn't be able to buy a new car.
d ...we'd spend our holidays with them there.
e ...I'd go round the world.
f ...you'd get there in three hours.

1 If you took the train, you'd get there in three hours.

7 Write sentences saying what you would do if...

1 you found £50 in the street.
2 someone asked you to lend them £50.
3 you spoke several languages.
4 your car broke down on the motorway.
5 you met an old friend on the way to work/school.
6 you had an important exam to take.

8 Rewrite these sentences with the past simple or the past perfect form of the verbs in brackets.

1 After I (leave) the office, I (go) straight home.
2 When they (arrive) at the station, they (miss) the train.
3 She (walk) slowly because she (hurt) her ankle.
4 He (write) to me after he (have) the accident.
5 She (come) downstairs when she (change) her clothes.
6 She (be) late because she (get) lost.

9 Answer the questions with *because* or *after* and the words in brackets.

1 Why did he feel ill?   (eat too much)
2 When did she leave?   (pay the bill)
3 When did he buy the book?   (read the reviews)
4 When did she learn French?   (learn English)
5 Why did he get lost?   (forget the way)
6 When did she give the newspaper back?   (read it)

1 Because he had eaten too much.

SOUNDS

1 Say these words aloud. Is the underlined sound /w/ or /r/? Put the words in two columns.

<u>w</u>orld   <u>r</u>eturn   <u>w</u>eeks   <u>r</u>eligion   <u>w</u>ash   <u>r</u>elative
<u>w</u>ater   <u>r</u>epair   <u>w</u>ashing   <u>wr</u>apping   <u>w</u>ear   <u>r</u>emind

Listen and check.

2 Say these words aloud. Is the underlined sound /ɔː/ or /aʊ/? Put the words in two columns.

p<u>oor</u>   p<u>ow</u>er   l<u>aw</u>   t<u>ow</u>er   s<u>our</u>   h<u>our</u>   m<u>ore</u>   m<u>oor</u>
r<u>oar</u>   s<u>aw</u>   b<u>ore</u>   fl<u>ow</u>er   s<u>ure</u>   p<u>our</u>   f<u>our</u>   w<u>ar</u>

Listen and check.

3 Listen and underline the stressed words in this passage.

> A man from Leipzig in Germany, who had wanted to be a rock musician for many years, decided to go to England. When he had packed his bags into his old Trabant car, he set off for Liverpool, home of the Beatles, to make his name as a musician.

4 Write the stressed words on a separate piece of paper. Now turn to Communication activity 8 on page 99.

SPEAKING

1 Work in groups of two or three. In what situations would you...

– ask someone how long they are going to stay
– tell a friend the truth if it would hurt them
– ask someone to leave
– tell someone a lie
– lose your temper with a friend
– ask someone to be quiet

2 Work in groups and discuss your answers to 1. Choose a situation which amuses, surprises or shocks you and write a dialogue.

A: *Hello, lovely to see you! When are you leaving?*
B: *Well, if it's all right with you, I'll stay for about six months.*
A: *What? Er, I mean, how nice...!*

3 Act out your dialogues to the rest of the class.

Communication activities

1 *Lesson 15*
Listening and speaking, activity 2

Student A: 🔲 Listen and find out what Karen has for breakfast and what Pat has for lunch.

When the recording stops, turn back to page 35.

2 *Lesson 16*
Listening and speaking, activity 1

Student A: 🔲 Listen and find out what type of entertainment karaoke is, and what type of music they play. Find out where they perform tango and the reasons why people enjoy it.

When the recording stops, turn back to page 38.

3 *Progress check 16 – 20*
Speaking and writing

Student B: Dictate these sentences to *Student A* in turn. Write down the sentences *Student A* dictates.

1 _____

2 So when a man finally got tickets he was surprised to find an empty seat between him and the next person, a woman dressed in black.

3 _____

4 The woman replied, 'Yes, we bought them some months ago but then my husband died.'

5 _____

6 The woman said, 'Well, they're all at the funeral.'

4 *Lesson 35*
Listening and speaking, activity 1

Student B: 🔲 Listen and put a tick (✓) by Stephen's answers to these questions.

– What time does he have lunch and dinner?
   midday and 5pm ☐   12.30pm and 6pm ☐
   1pm and 7pm ☐

– Does he use a napkin? If so, where does he put it?
   tucked under his chin ☐   tied round his neck ☐
   on his lap ☐

– Where does he put his hands when he's at the table but not eating?
   on the table ☐   he puts his elbows on the table ☐
   on his lap ☐

– When does he usually drink coffee and tea?

	morning	afternoon	any time
coffee			
tea			

– Does he have soup in the summer?
   often ☐   sometimes ☐   never ☐

When the recording stops, turn back to page 83.

5 *Lesson 23*
Vocabulary and listening, activity 3

Group A: 🔲 Listen to Barry and find the answers to these questions.

1 When is Australia Day?
2 What does the Melbourne Cup celebrate?
3 In what month is the Melbourne Cup?
4 What do people do before the race?
5 Who takes the day off work?

When the recording stops, check you have all got the same answers. Then turn back to page 54.

6 *Lesson 35*
Listening and speaking, activity 1

Student A: 🔲 Listen and put a tick (✓) by Stephen's answers to these questions.

– What does he say at the start of a meal?
'Enjoy your meal!' ☐   'Cheers!' ☐   nothing ☐

– Does he usually use a knife and fork? If so, which hands does he hold them in?
fork in the left hand, knife in the right ☐
he cuts with the fork in his left hand and the knife in his right, then puts the fork in his right hand to eat ☐
he doesn't use a knife, he holds the fork in his left hand ☐

– Where does he put his knife and fork when he has finished a meal?
together in the centre of plate, with the handles pointing towards him ☐
together on the plate, slightly sideways ☐
on the plate in a V-shape ☐

– What food does he often eat with his fingers at the dining table?
chicken ☐   bread ☐   cheese ☐   cake ☐
chocolate ☐   fruit ☐   pie ☐   lettuce ☐   chips ☐

– What does he say when someone raises their glass?
'Cheers!' ☐   'Health!' ☐   nothing ☐

When the recording stops, turn back to page 83.

7 *Lesson 14*
Speaking, activity 2

Student A: Give *Student B* directions from places in column 1 to places in column 2. Tell them where you start from but not where you're going to.

1	2
Trinity College	Abbey Theatre
the Custom House	the Bank of Ireland
the Irish Parliament	Pearse Station

Change round when you're ready.

8 *Progress check 36 – 40*
Sounds, activity 4

Without looking at the passage, write it out in full. Use the stressed words to help you.

Now look back at the passage and check your version.

9 *Lesson 11*
Reading, activity 3

Mostly 'a'
You're extremely ambitious. You're never satisfied with your life and you're always trying to improve things. Try to relax and take things easy!

Mostly 'b'
You're fairly ambitious. You are very aware that life has much to offer, but you don't feel you can achieve very much. Keep trying but don't make yourself unhappy.

Mostly 'c'
You're so unambitious, you don't even know the meaning of the word. Look it up in a dictionary, if you can be bothered.

10 *Progress check 6 – 10*
Writing, activity 2

Read the passage and complete your version of the story.

> **The least successful annual conference was in 1985 when the Association of British Travel Agents went to Sorrento. The flight from Gatwick Airport to Naples was delayed because of fog. Many people were ill because of something they ate. The organisers asked the Minister of Development to give a speech to the delegates in the forum at Pompeii, so a local travel agent decided to drop 3,500 roses as a gesture of friendship. As the Minister began his speech, a plane flew low over the audience and dropped the roses, but the flowers landed outside the forum. No one heard the speech because the engine noise was so loud. Minutes later it returned with some more roses, and missed again. Five times it passed over Pompeii, and each time the roses landed on Mount Vesuvius. The last time it flew over the delegates so low that it forced them to lie on the ground.**
>
> Adapted from *The return of heroic failures*, by Stephen Pile

11 *Lesson 16*
Listening and speaking, activity 1

Student C: 🔊 Listen and find out who performs karaoke. Find out what type of entertainment tango is and what type of music they play.

When the recording stops, turn back to page 38.

12 *Lesson 15*
Listening and speaking, activity 2

Student B: 🔊 Listen and find out what Karen has for lunch and what Pat has for dinner.

When the recording stops, turn back to page 35.

13 *Lesson 18*
Vocabulary, activity 2

The words you chose to describe the first person show the kind of person you'd like to be. The words you chose for the second person show how you think other people see you. The words you use to describe the third person show the real you, your true character!

14 *Lesson 35*
Listening and speaking, activity 1

Student C: 🔊 Listen and put a tick (✓) by Stephen's answers to these questions.

– How long does a typical lunch or dinner last?
  15 minutes ☐   30 minutes ☐   45 minutes ☐

– At which meals does he eat the following food?

	melon	pasta	fish	steak
breakfast				
lunch or dinner				
never				

– Does he eat cake with a fork or a spoon?
  fork ☐   spoon ☐   fork or spoon ☐   neither ☐

– When can he smoke during a meal?
  before, during and after ☐
  before and after ☐   never ☐

– Does he eat salad in the winter?
  often ☐   sometimes ☐   never ☐

When the recording stops, turn back to page 83.

15 *Lesson 14*
Speaking, activity 2

Student B: Give *Student A* directions from places in column 1 to places in column 2. Tell them where you start from but not where you're going to.

1	2
St Stephen's Green	the National Library
Merrion Square	the Custom House
Lower Baggott Street	Trinity College

Change round when you're ready.

16 *Progress check 16 – 20*
Speaking and writing

Student A: Dictate these sentences to *Student B* in turn. Write down the sentences *Student B* dictates.

1 *The Phantom of the Opera* was one of London's most popular musicals and it was difficult to reserve seats.
2 _____
3 He said, 'It took me a long time to get tickets for this show.'
4 _____
5 The man said, 'I'm so sorry. But why didn't you ask a friend or a relative to come with you?'
6 _____

17 *Lesson 40*
Speaking and writing, activity 3

The police stopped him and asked him where he had come from and what was wrong with the car. Unfortunately, he had learnt his English from Beatles songs, so he couldn't understand the police.

The police couldn't understand him either, and when they discovered that he had no money, they arrested him and sent him back to Germany. You can still see the rusty Trabant in the bushes, next to the M25.

18 *Lesson 23*
Vocabulary and listening, activity 3

Group B: 🔲 Listen to Barry and find the answers to these questions.

1 What does Australia Day celebrate?
2 On what day of the week is the Melbourne Cup and at what time of day?
3 How old is the race?
4 How long does it last?
5 Who is interested in the race?

When the recording stops, check you have all got the same answers. Then turn back to page 54.

19 *Lesson 16*
Listening and speaking, activity 1

Student B: 🔲 Listen and find out where the perform karaoke and the reasons why people enjoy it. Find out who performs tango.

When the recording stops, turn back to page 38.

20 *Lesson 15*
Listening and speaking, activity 2

Student C: 🔲 Listen and find out what Pat has for breakfast and what Karen has for dinner.

When the recording stops, turn back to page 35.

21 *Lesson 13*
Listening, activity 2

Caracas
GUYANA
VENEZUELA SURINAM
FRENCH GUIANA
Cali Bogotá
COLOMBIA
Quito
EQUADOR
Amazon
PERU
Recife
Lima
BRAZIL
L. Titicaca
Brasília
Machu Picchu La Paz
Belo Horizonte
BOLIVIA
Rio de Janeiro
PARAGUAY
São Paulo
CHILE Asunción
Mendoza Córdoba
Valparaíso URUGUAY
Santiago Buenos Aires
ARGENTINA Montevideo
Bahía Blanca
Patagonia
PACIFIC OCEAN
SOUTH ATLANTIC OCEAN

Grammar review

CONTENTS

Present simple

Form

You use the contracted form in spoken and informal written English.

Be

Affirmative	Negative
I'm (I am)	I'm not (am not)
you	you
we 're (are)	we aren't (are not)
they	they
he	he
she 's (is)	she isn't (is not)
it	it

Questions	Short answers
Am I?	Yes, I am.
	No, I'm not.
Are you/we/they?	Yes, you/we/they are.
	No, you/we/they're not.
Is he/she/it?	Yes, he/she/it is.
	No, he/she/it isn't.

Have

Affirmative	Negative
I	I
you have	you haven't (have not)
we	we
they	they
he	he
she has	she hasn't (has not)
it	it

Questions	Short answers
Have I/you/we/they?	Yes, I/you/we/they have.
	No, I/you/we/they haven't.
Has he/she/it?	Yes, he/she/it has.
	No, he/she/it hasn't.

Regular verbs

Affirmative	Negative
I	I
you work	you don't (do not) work
we	we
they	they
he	he
she works	she doesn't (does not) work
it	it

Questions	Short answers
Do I/you/we/they work?	Yes, I/you/we/they do.
	No, I/you/we/they don't.
Does he/she/it work?	Yes, he/she/it does.
	No, he/she/it doesn't.

Question words with *is/are*
What's your name?   Where are your parents?

Question words with *does/do*
What do you do?   Where does he live?

Present simple: third person singular
(See Lesson 2)

You add *-s* to most verbs.
takes, gets

You add *-es* to *do, go* and verbs ending in
-ch, -ss, -sh and *-x.*
goes, does, watches, finishes

You add *-ies* to verbs ending in *-y.*
carries, tries

Use
You use the present simple:

● to talk about customs and habits. (See Lesson 1)
In my country men go to restaurants on their own.

● to talk about routine activities. (See Lesson 2)
He gets up at 6.30.

● to talk about a habit. (See Lesson 5)
He smokes twenty cigarettes a day.

● to talk about a personal characteristic. (See Lesson 5)
She plays the piano.

● to talk about a general truth. (See Lesson 5)
You change money in a bank.

Present continuous

Form
You form the present continuous with *be* + present participle (*-ing*). You use the contracted form in spoken and informal written English.

Affirmative	Negative
I'm (am) working	I'm not (am not) working
you	you
we 're (are) working	we aren't (are not) working
they	they
he	he
she 's (is) working	she isn't (is not) working
it	it

Questions	Short answers
Am I working?	Yes, I am.
	No, I'm not.
Are you/we/they working?	Yes, you/we/they are.
	No, you/we/they aren't.
Is he/she/it working?	Yes, he/she/it is.
	No, he/she/it isn't.

Question words
What are you doing?   Why are you laughing?

Present participle (*-ing*) endings

You form the present participle of most verbs by adding *-ing.*
go – going, visit – visiting

You add *-ing* to verbs ending in *-e.*
make – making, have – having

You double the final consonant of verbs of one syllable ending in a vowel and a consonant, and add *-ing.*
get – getting, shop – shopping

You add *-ing* to verbs ending in a vowel and *-y* or *-w.*
draw – drawing, play – playing

You don't usually use these verbs in the continuous form.
believe feel hear know like see smell sound taste think understand want

Use
You use the present continuous to say what is happening now or around now. There is an idea that the action or state is temporary. (See Lesson 5)
It's raining.   I'm learning English.

Past simple

Form

You use the contracted form in spoken and informal written English.

Be

Affirmative		Negative	
I		I	
he	was	he	wasn't (was not)
she		she	
it		it	
you		you	
we	were	we	weren't (were not)
they		they	

Have

Affirmative		Negative	
I		I	
you		you	
we		we	
they	had	they	didn't (did not) have
he		he	
she		she	
it		it	

Regular verbs

Affirmative		Negative	
I		I	
you		you	
we		we	
they	worked	they	didn't work
he		he	
she		she	
it		it	

Questions	Short answers
Did I/you/we/they work?	Yes, I/you/we/they did.
he/she/it	he/she/it
	No, I/you/we/they didn't.
	he/she/it

Question words
What did you do?   Why did you leave?

Past simple endings

You add -*ed* to most regular verbs.
walk – walked, watch – watched

You add -*d* to verbs ending in -*e*.
close – closed, continue – continued

You double the consonant and add -*ed* to verbs ending in a vowel and a consonant.
stop – stopped, plan – planned

You drop the -*y* and add -*ied* to verbs ending in -*y*.
study – studied, try – tried

You add -*ed* to verbs ending in a vowel + -*y*.
play – played, annoy – annoyed

Irregular verbs

There are many verbs which have an irregular past simple. For a list of the irregular verbs which appear in **Reward Pre-intermediate**, see page 112.

Pronunciation of past simple endings

/t/ *finished, liked, walked*
/d/ *continued, lived, stayed*
/ɪd/ *decided, started, visited*

Expressions of past time
(See Lesson 8)

yesterday, the day before yesterday, last weekend, last night, last month, last year

Use
You use the past simple:

● to talk about a past action or event that is finished. (See Lessons 6, 7 and 8)
We walked towards each other.

Future simple (*will*)

Form

You form the future simple with *will* + infinitive. You use the contracted form in spoken and informal written English.

Affirmative		Negative	
I		I	
you		you	
we		you	
they	'll (will) work	they	won't (will not) work
he		he	
she		she	
it		it	

Questions	Short answers
Will I/you/we/they work?	Yes, I/you/we/they will.
he/she/it/	he/she/it/
	No, I/you/we/they won't.
	he/she/it/

Question words
What will you do?   Where will you go?

Expressions of future time

tomorrow, tomorrow morning, tomorrow afternoon,
next week, next month, next year, in two days,
in three months, in five years

Use

You use the future simple:

- to make a prediction or express an opinion about the future. (See Lesson 12)
  I think most people will need English for their jobs.

- to talk about decisions you make at the moment of speaking. (See Lesson 13)
  I'll give you the money right now.

- to talk about things you are not sure will happen with probably and perhaps. (See Lesson 13)
  He'll probably spend three weeks there.
  Perhaps he'll stay two days in Rio.

- to offer to do something. (See Lessons 13 and 30)
  OK, I'll buy some food.

Present perfect simple

Form

You form the present perfect with *has/have* + past participle. You use the contracted form in spoken and informal written English.

Affirmative		Negative	
I		I	
you	've (have) worked	you	haven't (have not) worked
we		we	
they		they	
he		he	
she	's (has) worked	she	hasn't (has not) worked
it		it	

Questions	Short answers
Have I/you/we/they worked?	Yes, I/you/we/they have.
	No, I/you/we/they haven't.
Has he/she/it worked?	Yes, he/she/it have.
	No, he/she/it hasn't.

Past participles

All regular and some irregular verbs have past participles which are the same as their past simple form.
Regular: *move – moved, finish – finished, visit – visited*
Irregular: *leave – left, find – found, buy – bought*

Some irregular verbs have past participles which are not the same as the past simple form.
go – went – gone   be – was/were – been
drink – drank – drunk   ring – rang – rung

For a list of the past participles of the irregular verbs which appear in **Reward Pre-intermediate**, see page 112.

Been and gone

He's been to America = He's been there and he's back here now.
He's gone to America = He's still there.

Use

You use the present perfect:

- to talk about past experiences. You often use it with *ever* and *never*. (See Lesson 21)

  Have you ever stayed in hospital?
  I've had food poisoning several times.
  I've never broken my leg.

- to talk about a past action which has a result in the present. It is not important when the action happened. You often use it to describe changes. (See Lesson 22)

  She's got married.
  I've moved to a new flat.
  Have you found a new job?

  You often use *just* to emphasise that something has happened very recently.
  She's just had a baby.

- to talk about an action or state which began in the past and continues to the present. You use *for* to talk about the length of time. (See Lesson 23)
  I've been here for two hours.

  You use *since* to say when the action or state began.
  I've been here since 8 o'clock.

8 o'clock	Now – 10 o'clock
I arrived.	*I am still here.*

Remember that if you ask for and give more information about these experiences, actions or states, such as *when, how, why* and *how long*, you use the past simple.

When did you stay in hospital?   In 1975.
How long did you stay there?   A week.

Past continuous

Form

You form the past continuous with *was/were* + present participle. You use the contracted form in spoken and informal written English.

Affirmative	Negative
I	I
he was working	he wasn't (was not) working
she	she
it	it
you	you
we were working	we weren't (were not) working
they	they

Questions	Short answers
Was I/he/she/it working?	Yes, I/he/she/it was.
	No, I/he/she/it wasn't.
Were you/we/they working?	Yes, you/we/they were.
	No, you/we/they weren't.

You don't usually use these verbs in the continuous form.
believe feel hear know like see smell sound taste think understand want

Use

You use the past continuous:

● to talk about something that was in progress at a specific time in the past. (See Lesson 31)

What were you doing at nine o'clock yesterday morning? I was going to work.

8.30am	**9am**	**9.30am**
I left home. ———————————→		*I arrived at work.*

● to talk about something that was in progress at the time something else happened or interrupted it. You join the parts of the sentences with *when* and *while*. The verb in the *when* clause is usually in the past simple. (See Lessons 31 and 32)

I was watching television when there was a knock at the door.

The verb in the *while* clause is usually in the past continuous.
While they were visiting the Palace, one of the women had an idea.

Remember that you use *when* + past simple to describe two things which happened one after the other.
When the two women got closer, they saw some people in eighteenth-century clothes.

Past perfect

Form

You form the past perfect with *had* + past participle. You use the contracted form in spoken and informal written English.

Affirmative	Negative
I	I
you	you
we	we
they 'd (had) worked	they hadn't (had not) worked
he	he
she	she
it	it

Questions	Short answers
Had I/you/we/they worked?	Yes, I/you/we/they had.
he/she/it/	he/she/it/
	No, I/you/we/they hadn't.
	he/she/it/

Use

You use the past perfect:

● to talk about one past action that happened before another past action. You often use *after, when* and *because*, and you use the past simple for the second action. (See Lesson 40)

After he had decided to go to Britain, he packed his bags.

Earlier past	Past	Now
He decided to go to Britain.	*He packed his bags.*	→

When he had packed his bags into his car, he set off. Because he had spent so much, he had very little money.

Verb patterns

There are several possible patterns after certain verbs which involve *-ing* form verbs and infinitive constructions with or without *to*.

-ing form verbs

You can put an *-ing* form verb after certain verbs. (See Lesson 4)
I love walking.   She likes swimming. They hate lying on a beach.

Remember that *would like to do something* refers to an activity at a specific time in the future.
I'd like to go to the cinema next Saturday.

Try not to confuse it with *like doing something* which refers to an activity you enjoy all the time.
I like going to the cinema. I go most weekends.

Use

You use an *-ing* form verb:

- to describe the purpose of something after *to be for*.
  (See Lesson 25)
  A cassette player is for playing cassettes.
- to ask people to do things after *would you mind*.
  (See Lesson 30)
  Would you mind lending it to me?

To + infinitive

You can put *to* + infinitive after many verbs. Here are some of them:
agree decide go have hope learn leave like need offer start try want

Use

You use *to* + infinitive:

- with *(be) going to* and *would like to*. (See below)
- to describe the purpose of something. (See Lesson 25)

You use a cassette player to play cassettes.

Going to

You use *(be) going to*:

- to talk about future intentions or plans.
  (See Lessons 11 and 25)
  I'm going to be a doctor. (I'm studying medicine.)
- to talk about things which are arranged and sure to happen with *(be) going to*. (See Lesson 13)
  I'm going to visit South America. I've got my ticket.

You often use the present continuous and not *going to* with *come* and *go*.
Are you coming tonight?
NOT ~~Are you going to come tonight?~~
He's going to South America.
NOT ~~He's going to go to South America.~~

Would like to

You use *would like to*:

- to talk about ambitions, hopes or preferences.
  (See Lesson 11)
  I'd like to speak English fluently.

Modal verbs

The following verbs are modal verbs.
can could may might must should will would

Form
Modal verbs:

- have the same form for all persons.
  I must leave. He must be quiet.

- don't take the auxiliary *do* in questions and negatives.
  Can you drive? You mustn't lean out of the window.
- take an infinitve without *to*.
  I can type. You should see a doctor.

Use

You use *can*:

- to express general ability, something you are able to do on most occasions. (See Lesson 27)
  I can swim a hundred metres.
- to say what you're allowed to do or what it is possible to do. (See Lesson 28)
  You can cross when the light is green.
- to ask for permission. (See Lesson 30)
  Can I smoke?
- to ask people to do things. (See Lesson 30)
  Can you speak louder, please?

Can is a little less formal than *could*.

You use *can't*:

- to say what you're not allowed to do or what it is not possible to do. (See Lesson 28)
  You can't cross when the light is red.

  You can also use *mustn't*. (See Lesson 26)
  You mustn't cross when the light is red.

You use *could*:

- to express general ability in the past. (See Lesson 27)
  When I was five, I could swim but I couldn't write my own name.
- to ask for permission. (See Lesson 30)
  Could I leave now?
- to ask people to do things. (See Lesson 30)
  Could you help me?

  Could is a little more formal than *can*.

You use *may*:

- to ask for permission. (See Lesson 30)
  May I call you Esther?
- to talk about possible future events. (See Lesson 36)
  It may rain tomorrow.

  May has almost the same meaning as *might*.

You use *mght*:

- to talk about possible future events. (See Lesson 36)
  It might rain tomorrow.

  You don't usually use *might* in questions.

You use *must*:

- to talk about something you are obliged to do. The obligation usually comes from the speaker and it can express a moral obligation, necessity, strong advice or a strong suggestion. (See Lesson 26)
  It's late. I must go now. You really must stop smoking.

You often use *must* to talk about safety instructions.
You must fasten your seatbelt.

Have to has almost the same meaning as *must* but the obligation comes from a third person. You often use it to talk about rules.
The government says you have to do military service.

You use *mustn't*:

● To talk about something you're not allowed to do. (See Lesson 26)
You mustn't smoke here.

You can also use *can't*.
You can't smoke here.

You use *should*:

● to give advice. It can also express a mild obligation or the opinion of the speaker. (See Lesson 29)
You should take some exercise. You shouldn't smoke.

● *Ought(n't) to* has the same meaning as *should(n't)*.
You ought to take some exercise.

For the uses of *will* see *Future simple (will)*

You use *would*:

● to ask for permission with *mind if*. (See Lesson 30)
Would you mind if I borrowed it?

● to ask people to do things with *mind + -ing* form verbs. (See Lesson 30)
Would you mind lending it to me?

● to talk about the consequences of an imaginary situation. (See Lesson 38)
I'd go to a small town in France.

Conditionals

First conditional

Form

You form the first conditional with *if* + present simple, *will* + infinitive.
If you touch it, you'll get a shock.

Use

You use the first conditional to talk about a likely situation and to describe its result. You talk about the likely situation with *if* + present simple. You describe the result with *will* or *won't*. (See Lesson 37)
If you give it to me, I'll let you go.
If it doesn't stop rising, we won't be able to escape.

You often use the contracted form in speech and informal writing. The *if* clause can come at the beginning or at the end of the sentence.

Second conditional

Form

You form the second conditional with *if* + past simple, *would* + infinitive.
If I found some money in the street, I'd take it to the police.

You can also use the past continuous in the *if* clause.
If it was raining, I'd take an umbrella.

Use

You use the second conditional to talk about an imaginary or unlikely situation and its result. You talk about the imaginary or unlikely situation with *if* + past tense. You describe the result with *would* or *wouldn't*. (See Lesson 39)
If I had a lot of money, I would give some away.
If a stranger asked me for money, I wouldn't give him any.

You often use the contracted form in speech and informal writing. The *if* clause can come at the beginning or at the end of the sentence. It is still common to see *were* and not *was* in the *if* clause.
If I were you, I'd go home

Present simple passive

Form

You form the present simple passive with *am/is/are* + past participle.
A splendid feast is prepared.
Shops are filled with symbols of death.

Use

You use the passive to focus on the object of the sentence. You can use it when you don't know who or what does something. The object of an active sentence becomes the subject of a passive sentence. (See Lesson 33)
A bowl of water is placed on the table.

If you are more interested in the object but you know who or what does something, you use *by*.
A special kind of bread is baked by the men.

Have got

Form

You use the contracted form in spoken and informal written English.

Affirmative		**Negative**	
I		I	
you	've (have) got	you	haven't (have not) got
we		we	
they		they	
he		he	
she	's (has) got	she	hasn't (has not) got
it		it	

Questions	Short answers
Have I/you/we/they got?	Yes, I/you/we/they have.
	No, I/you/we/they haven't.
Has he/she/it got?	Yes, he/she/it has.
	No, he/she/it hasn't.

Use

You use *have got* to talk about facilities, possession or relationship. (See Lesson 10)
I've got a new car.

You don't use *have got* to talk about a habit or routine.
I often have lunch out. NOT I often have got lunch out.

You don't usually use *have got* in the past. You use the past simple of *have*.
I had a headache yesterday. NOT I had got a headache.

Questions

You can form questions in two ways:

● with a question word such as *who, what, which, where, how, why*.
What's your name?

● without a question word.
Are you English?

You can put a noun after *what* and *which*. (See Lesson 1)
What time is it?   Which road will you take?

You often say *what* to give the idea that there is more choice.
What books have you read lately?

You can put an adjective or an adverb after *how*.
(See Lessons 15 and 20)
How much is it?   How long does it take by car?
How fast can you drive?

You can use *who, what* or *which* as pronouns to ask about the subject of the sentence. You don't use *do* or *did*.
(See Lessons 1 and 7)
What's your first name?
Who organised the first package trip?

You can use *who, what* or *which* and other question words to ask about the object of the sentence. You use *do* or *did*.
(See Lessons 1 and 7)
Who did he take on the first package trip?

Defining relative clauses

● You use a defining relative clause to define people, things and places. The information in the defining relative clause is important for the sense of the sentence. (See Lesson 24)

You use *who* for people.
A druggist is someone who sells medicine in a shop.

You use *which* for things.
A subway is a railway which runs under the ground.

You often use *that* instead of *who* or *which*.
A druggist is someone that sells medicine in a shop.
A subway is a railway that runs under the ground.

You use *where* for places.
A parking lot is a place where you park your car.

You can use *that* after a superlative adjective instead of *who, which* or *where*. (See Lesson 19)
Who is the nicest person that you know?
What is the most expensive thing that you own?

Remember that there is no comma before a defining relative clause. In speech there is no pause.

Articles

You can find the main uses of articles in Lesson 3. Here are some extra details.

You use *an* for nouns which begin with a vowel.
an armchair

You use *one* if you want to emphasise the number.
One hundred and twenty-two.

Before vowels, you pronounce *the* /ði:/

You do not use the definite article with parts of the body.
You use a possessive adjective.
I'm washing my hair.

Plurals

You can find the main rules for forming plurals in Lesson 3.

Possessives

You can find the main uses of the possessive *'s* in Lesson 9.
You can find a list of possessive adjectives in Lesson 9.

Expressions of quantity

Countable and uncountable nouns

Countable nouns have both a singular and a plural form.
(See Lesson 15)
an apple – some apples, a melon – some melons,
a potato – some potatoes, a cup – (not) many cups,
a biscuit – a few biscuits

Uncountable nouns do not usually have a plural form.
some wine, some cheese, some fruit, (not) much meat,
a little coffee

If you talk about different kinds of uncountable nouns, they become countable.
Beaujolais and Bordeaux are both French wines.

Expressions with countable or uncountable nouns

You can put countable or uncountable nouns with these expressions of quantity.
lots of apples, lots of cheese, hardly any apples, harldy any cheese, quite a lot of fruit, quite a lot of potatoes

Some and any
(See Lesson 15)

Affirmative   *There's some milk in the fridge.*
                  There are some apples on the table.

Negative   *I haven't got any brothers.*
                  There isn't any cheese.

Questions

You usually use *any* for questions.
Is there any sugar?

You can use *some* in questions when you are making an offer or a request, and you expect the answer to be *yes*.
Would you like some tea?
Can I have some sugar, please?

Much and many

You use *many* with countable nouns and *much* with uncountable nouns. (See Lesson 15)
How many eggs would you like?
How much butter do you need?

Too much/many, not enough, fewer, less and more

You can put a countable noun in the plural after *too many, not enough* and *fewer*.
There are too many people.
There aren't enough clean rivers.
In Britain there are fewer men than women.

You can put an uncountable noun after *too much, not enough, more* and *less*.
There's too much noise.   There isn't enough farmland.
There's more pollution.

You can put an adjective after *too* or between *not* and *enough*.
The sea is too polluted.   The air isn't clean enough.

Making comparisons
Comparative and superlative adjectives

Form

You add *-er* to most adjectives for the comparative form, and *-est* for the superlative form. (See Lesson 18)
cold colder coldest   cheap cheaper cheapest

You add *-r* to adjectives ending in *-e* for the comparative form, and *-st* for the superlative form.
large larger largest   fine finer finest

You add *-ier* to adjectives ending in *-y* for the comparative form, and *-iest* for the superlative form.
happy happier happiest
friendly friendlier friendliest

You double the last letter of adjectives ending in *-g, -t,* or *-n.*
hot hotter hottest   thin thinner thinnest

You use *more* for the comparative form and *most* for the superlative form of longer adjectives.
expensive more expensive most expensive
important more important most important

Some adjectives have irregular comparative and superlative forms.
good better best   bad worse worst

More than, less than, as ... as

● You put *than* before the object of the comparison. (See Lesson 19)
Children wear more casual clothes than their parents.

● You use *less ... than* to change the focus of the comparison.
Parents wear less casual clothes than their children.

● You can put *much* before the comparative adjective, *more* or *less* to emphasise it.
They're much less formal than they were.

● You use *as ... as* to show something is the same.
They're as casual as teenagers are all over the world.

● You use *not as ... as* to show something is different.
Dresses are not as popular as in Western countries.

But, however, although

You use *but, however,* and *although* to make a comparison which focuses on a difference. (See Lesson 35)

You put *but* at the beginning of a sentence or to join two sentences.

We drink coffee in the morning. But we don't drink it in the afternoon.
We drink coffee in the morning but we don't drink it in the afternoon.

You use *although* at the beginning of a subordinate clause. You need to separate the subordinate and the main clause with a comma.
We usually have dinner at six, although some people have dinner later.

You can put the subordinate clause at the beginning or at the end of the sentence.
Although we usually have dinner at six, some people have dinner later.

You put *however* at the beginning of a sentence. You put a comma after it.
We drink coffee in the morning. However, we don't drink it in the afternoon.

So, because

- You can join two sentences with *so* to describe a consequence.
  She often took the plane so she didn't look at the safety instructions.

- You can join the same two sentences with *because* to describe a reason.
  She didn't look at the safety instructions because she often took the plane.

Prepositions of place

(See Lesson 14)

Prepositions of time and place: *in, at, on, to*

(See Lesson 16)

Use

You use *in*:

- with times of the day: *in the morning, in the afternoon*
- with months of the year: *in March, in September*
- with years: *in 1996, in 1872*
- with places: *in London, in France, in the cinema*

You use *at*:

- with times of the day: *at night, at seven o'clock*
- with certain expressions of time: *at the weekend*
- with places: *at the theatre, at the stadium*

You use *on*:

- with days, dates: *on Friday, on 15th July*

You use *to*:

- with places: *Let's go to London.*

Adverbs of frequency

Use

You use an adverb of frequency to say how often things happen. (See Lesson 1)
They always take their shoes off.
We usually take chocolates or flowers.
We often wear jeans and sweaters.
We sometimes arrive about fifteen minutes before.
We never ask personal questions.

Pronunciation guide

/ɑː/	p<u>ar</u>k	/b/	<u>b</u>uy
/æ/	h<u>a</u>t	/d/	<u>d</u>ay
/aɪ/	m<u>y</u>	/f/	<u>f</u>ree
/aʊ/	h<u>ow</u>	/g/	<u>g</u>ive
/e/	t<u>e</u>n	/h/	<u>h</u>ouse
/eɪ/	b<u>ay</u>	/j/	<u>y</u>ou
/eə/	<u>the</u>re	/k/	<u>c</u>at
/ɪ/	s<u>i</u>t	/l/	<u>l</u>ook
/iː/	m<u>e</u>	/m/	<u>m</u>ean
/ɪə/	b<u>ee</u>r	/n/	<u>n</u>ice
/ɒ/	wh<u>a</u>t	/p/	<u>p</u>aper
/əʊ/	n<u>o</u>	/r/	<u>r</u>ain
/ɔː/	m<u>ore</u>	/s/	<u>s</u>ad
/ɔɪ/	t<u>oy</u>	/t/	<u>t</u>ime
/ʊ/	t<u>oo</u>k	/v/	<u>v</u>erb
/uː/	s<u>oo</u>n	/w/	<u>w</u>ine
/ʊə/	t<u>ou</u>r	/z/	<u>z</u>oo
/ɜː/	s<u>ir</u>	/ʃ/	<u>sh</u>irt
/ʌ/	s<u>u</u>n	/ʒ/	lei<u>s</u>ure
/ə/	bett<u>er</u>	/ŋ/	si<u>ng</u>
		/tʃ/	chur<u>ch</u>
		/θ/	<u>th</u>ank
		/ð/	<u>th</u>en
		/dʒ/	<u>j</u>acket

Irregular Verbs

Verbs with the same infinitive, past simple and past participle

cost	cost	cost
cut	cut	cut
hit	hit	hit
let	let	let
put	put	put
read /riːd/	read /red/	read /red/
set	set	set
shut	shut	shut

Verbs with the same past simple and past participle, but a different infinitive

bring	brought	brought
build	built	built
burn	burnt/burned	burnt/burned
buy	bought	bought
catch	caught	caught
feel	felt	felt
find	found	found
get	got	got
have	had	had
hear	heard	heard
hold	held	held
keep	kept	kept
learn	learnt/learned	learnt/learned
leave	left	left
lend	lent	lent
light	lit/lighted	lit/lighted
lose	lost	lost
make	made	made
mean	meant	meant
meet	met	met
pay	paid	paid
say	said	said
sell	sold	sold
send	sent	sent
sit	sat	sat
sleep	slept	slept
smell	smelt/smelled	smelt/smelled
spell	spelt/spelled	spelt/spelled
spend	spent	spent
stand	stood	stood
teach	taught	taught
understand	understood	understood
win	won	won

Verbs with same infinitive and past participle but a different past simple

become	became	become
come	came	come
run	ran	run

Verbs with a different infinitive, past simple and past participle

be	was/were	been
begin	began	begun
break	broke	broken
choose	chose	chosen
do	did	done
draw	drew	drawn
drink	drank	drunk
drive	drove	driven
eat	ate	eaten
fall	fell	fallen
fly	flew	flown
forget	forgot	forgotten
give	gave	given
go	went	gone
grow	grew	grown
know	knew	known
lie	lay	lain
ring	rang	rung
rise	rose	risen
see	saw	seen
show	showed	shown
sing	sang	sung
speak	spoke	spoken
swim	swam	swum
take	took	taken
throw	threw	thrown
wake	woke	woken
wear	wore	worn
write	wrote	written

Tapescripts

Lesson 1   **Speaking and listening, activity 2**

Situation 1

PATRICIA   Thierry! How lovely to see you. You do look well.
THIERRY   How are you Patricia?
PATRICIA   We're all very well, really well. Come in, come in.
THIERRY   Patricia, this is my friend Rosario Rodriguez.
PATRICIA   Hello, Rosario, how do you do. I'm very pleased to meet you.
ROSARIO   How do you do.
PATRICIA   Do come in. We're in the back room, come on through. Now, when was the last time we got together?...

Situation 2

WOMAN   Excuse me!
WAITER   Yes, can I help you?
WOMAN   Yes, I'd like a Coke, please.
WAITER   Certainly. With ice and lemon?
WOMAN   Sorry, I don't understand.
WAITER   Would you like ice and lemon in your Coke?
WOMAN   Oh, yes, please.
WAITER   Here you are.
WOMAN   Thank you. How much is this?
WAITER   Ninety pence.

Lesson 1   **Speaking and listening, activity 3**

1   What's your first name?
2   How old are you?
3   How much do you earn?
4   Where do you live?
5   Are you married?
6   What do you do?
7   Do you have any brothers and sisters?
8   Where do you come from?

Lesson 2   **Listening, activity 1**

Speaker 1

JO-ANN   I get up around seven o'clock and have breakfast. Then I have to leave home at about eight, I guess. It takes me about fifteen minutes to ride the subway downtown and start work.
Q   Do you like your job?
JO-ANN   It's OK. Some days it's fine, some days you get some real tricky customers.
Q   Do you get time off for lunch?
JO-ANN   Sure. I stop around twelve-thirty and then at one o'clock I take a walk in the park – just to get some fresh air, you know. If the weather's OK, I'll have a sandwich or something there.
Q   Mmm. Does it get busy in the afternoon?
JO-ANN   No more than in the morning, I guess. When the bank's open, people call at all times.
Q   So when do you leave work?
JO-ANN   I leave work at about six-fifteen in the evening, and at six-thirty I take the subway back home. I don't do much when I get home, except watch TV. I guess I go to bed at around eleven-thirty.

Speaker 2

GEORGE   I get up very early these days, much earlier than when I was at work, at about six o'clock, I guess.
Q   Why so early?
GEORGE   Oh, I like to see the sun rise, have a walk on the beach, you know. If it's hot, I go for a swim in the sea before breakfast.
Q   And what time do you have breakfast?
GEORGE   I get breakfast ready for about eight-fifteen and take it up to Hilary who's still in bed. Then we go shopping in the local mall, meet some friends, have lunch, that sort of thing.
Q   What time do you have lunch, then?
GEORGE   At about half-past twelve, I guess. And in the afternoon, I go for another walk, maybe play a little golf, have a swim, go down to the community centre to join Hilary.
Q   And what about in the evening?

GEORGE   Well, I meet my friends at about five-thirty and have a drink or two at the golf club. We talk for about an hour or so, then I go home for dinner at seven o'clock. Maybe we have friends round, maybe we go for dinner, it just depends. But most nights we're in bed by ten-thirty. It's a great life!

Lesson 2   **Sounds, activity 2**

/s/   takes sits asks talks
/z/   goes sings arrives offers has serves does
/ɪz/   finishes refuses washes watches

Lesson 5   **Listening and vocabulary, activity 2**

Conversation 1

WOMAN   Good morning.
CASHIER   Good morning. Can I help you?
WOMAN   I'd like two hundred pounds worth of Italian lira, please.
CASHIER   Traveller's cheques or cash?
WOMAN   Traveller's cheques.
CASHIER   Could I have your passport, please? Thank you. Would you sign and date them, please?
WOMAN   Here you are.
CASHIER   That makes four hundred and twenty one thousand lira. If you'd like to go over to the cash desk, the cashier will give you the money.
WOMAN   Thank you.
CASHIER   Thank you. Goodbye.

Conversation 2

WAITRESS   Good evening sir. A table for two?
MAN   Yes, but could we sit by the window, please?
WAITRESS   Certainly sir. Come this way. Here you are sir, madam. Can I take your coats?
WOMAN   Thank you.
WAITRESS   Here are the menus. Can I get you an aperitif?
MAN   Not for me, thank you.
WOMAN   No, thanks. I think we'll just have some wine with the meal, please.
WAITRESS   Could I suggest the lamb this evening? It's done in spices, it's a Lebanese dish, a speciality of the chef. I'm sure you'll like it. It comes with mixed vegetables.
MAN   Sounds lovely.
WOMAN   Well, I'd like to have a look at the menu anyway.
WAITRESS   Certainly, madam. I'll come back when you're ready to order.

Conversation 3

WOMAN   A return ticket to London, please.
OFFICIAL   Are you travelling after nine twenty?
WOMAN   Yes.
OFFICIAL   That's eleven pounds fifty. Would you like a travelcard?
WOMAN   What's that?
OFFICIAL   It gives you unlimited travel on the underground and the buses all day.
WOMAN   OK, I'll have a travelcard as well.
OFFICIAL   That'll be thirteen fifty. Thank you.
WOMAN   Thank you.

Conversation 4

MAN   What would you recommend for a cough?
CHEMIST   Well, I can give you some cough medicine, but if it's very bad, you ought to see a doctor.
MAN   No, it's not too bad. I thought I'd try something from the chemist's first.
CHEMIST   Have you got a headache and a temperature?
MAN   Yes.
CHEMIST   Well, I should try this medicine for your cough and these tablets for the headache and the temperature. They're a kind of aspirin which you can dissolve in a cup of hot water.
MAN   Thank you. How much is that?
CHEMIST   That'll be three pounds twenty, please.
MAN   Thank you.

Progress check 1 – 5   **Sounds, activity 1**

/s/   gets looks smokes wants
/z/   does goes leaves sings
/ɪz/   finishes refuses

Lesson 6   **Sounds, activity 2**

/t/   finished   walked   danced
/d/   continued   enjoyed   called
/id/   started   wanted   expected

Lesson 8   **Listening and speaking, activity 2**

A few years ago, I was going from London to Paris to join my husband and children. I checked in early and the assistant offered to change my ticket from tourist class to business class, which was very nice, and as she gave me my boarding pass, she told me there was no delay, because there was little air traffic. So all in all, I was in a very good mood. I went to the business class lounge and had some coffee, then they called the flight, right on time. I had a row of seats to myself because there weren't many other passengers and we all sat down, put on the seat belts and waited for take-off. The cabin crew went through the safety instructions and I thought, 'Why do they do this on every flight? It's only forty-five minutes to Paris.'

Then we took off, it was lovely morning so we had a fabulous view of the south coast as we flew towards France. But then the pilot made an announcement. 'I am sorry but we appear to have some technical difficulties, nothing to worry about, but we will land at an airport just outside Paris while we get things sorted out. 'He sounded very calm, but I knew that something was the matter. The flight attendants went through the cabin telling us to put on our safety belts. Then the captain came on again. 'Well, I'm afraid that the situation is more serious than we thought and we will be landing as soon as we reach the French coast.' So for the first time in my life I looked at the safety instructions and checked the emergency exits. In a very short time we landed, and as soon as the plane came to a stop, the attendants told us to get out as quickly as possible using the emergency chutes, and to get away from the plane. Nobody panicked, but as soon as we were all out of the plane, we ran as fast as we could. They told us later that they thought there was a bomb on board. It was only afterwards that I realised how frightened I was.

Lesson 8   **Listening and speaking, activity 5**

I arrived in this town quite late, about ten o'clock, I suppose, and looked for somewhere to stay. I didn't have a reservation anywhere, because I didn't expect to stay there. The main hotel was full, but the woman on the reception desk told me there was a small guest house down the road. I left the car where it was and I went down the road and in fact I walked past the guest house, because I saw there were no lights on. I knocked on the door – there was no bell, or anything – and a man opened the door. He was unshaven and wearing a very dirty vest, and the television was on. I asked if he had a room and without saying a word, he picked up a key from behind the desk and pushed a registration form at me. I wrote my name, but not my address. I don't know why, I felt there was something wrong. I saw that all the keys were on their hooks, so either all the other guests were out or I was the only person to stay there. I followed him up some very dirty stairs – there was no lift, of course – and he showed me a very dusty room. He gave me the key and closed the door. I sat down on the tiny single bed and wondered what to do. It was a horrible room with a washbasin, but no shower or toilet or anything. I decided I wanted to check out but I was frightened by the man downstairs. I ran downstairs and walked quickly past him, but he didn't look up. I quickly said, 'I left my suitcase in my car', and ran out of the door, up the street, and locked myself in the car – I was so scared. I slept in the car that night!

Lesson 9   **Vocabulary and listening, activity 2**

I have quite a large family, actually. My grandmother lives with my mother and father, and she's called Jacqueline. Then I have a brother who's called Raymond, who lives in Cassis, that's quite near Marseille, and a sister, called Chantal who lives in Marseille. My aunt is called Christine and my uncle is called Tony. Then, of course, there's my husband Vincent. His family comes from Montpellier, and that's where we live now. Oh, I nearly forgot, my mother is called Marie and my father is Georges. We're very close.

Lesson 10   **Vocabulary and listening, activity 2**

Q	What's Mestre like?
CATHERINE	Oh, it's not bad. I mean, people don't really visit Mestre at all, because they're on their way through to Venice. But, in a way, it's easier to live in Mestre, rather than Venice.
Q	Why's that?

CATHERINE	Well, it's cheaper, for one and it's a lot more convenient being on the mainland.
Q	Oh, right.
CATHERINE	But it's quite industrial, really, it's got a port and quite a few factories. I mean, it's not a pretty place to look at.
Q	What's the rest of the town like?
CATHERINE	Well, as far as culture goes, it hasn't got a cathedral or an art gallery. They're all in Venice. But the shops are very good, I think, and it's got a nice park as well.
Q	Any sports facilities?
CATHERINE	Unfortunately, nothing very big, no.
Q	Would you say the night-life is good?
CATHERINE	No, not really, I mean, it's got restaurants – the food can be very good – and a cinema, which we go to quite often. But if you want theatres or discos you have to go to Venice.

Lesson 10   **Vocabulary and listening, activity 5**

SPEAKER 1	It's quite cheap to live here, although in certain parts of town the accommodation is really very expensive. Everyday things in the shop are very cheap compared with England, although clothes can cost a lot.
SPEAKER 2	Oh, it's incredible. I mean, there are so many cars and lorries. You see, the railways system is not very extensive so everything has to go by road. But at rush hours in the centre of town, it's so busy.
SPEAKER 3	It's a very interesting place to visit, although I wouldn't call the buildings beautiful. But the palace is quite fascinating and well worth spending a day on. And then there are some wonderful churches and mosques.
SPEAKER 4	It's quite small, really, about one hundred thousand people, and the city centre is really very compact, although the suburbs stretch for about five kilometres in each direction.
SPEAKER 5	It rains all the time, it's quite famous for that. Even in the middle of summer you are never certain that the sun will shine all day, so it's difficult to make plans for any outdoor activities.
SPEAKER 6	Well, you can still walk the streets at night and not feel nervous that someone is going to attack you. But I think it's a good idea to be sensible and keep your money and valuables safe in your jacket. You never know, it can happen anywhere, can't it?

Lesson 12   **Listening, activity 2**

Q	Lynne, um... I'd like to ask you first, um... at what age do people start learning English these days?
LYNNE	Um... well, in many countries children start learning English when they go to school and then they complete their formal training later on but, I think in some countries they're starting to teach English to much younger children and I think this will become more and more common around the world.
GREG	Yeah, um... that's certainly true 'cause I know that er... in some countries they're even having English lessons for six-year-old children, um... so, er... they'll certainly be learning as soon as they start school, if not before.
Q	I see. And do you think that, um... English will soon be the universal language?
LYNNE	Oh, I think most adults already speak some English, um... even if it's only a word or two here and there, because, um... well, English is very common and very useful.
GREG	Mmm... I...
Q	What about you Greg?
GREG	Well, I was just going to say that, er... I think that's right. Because, if you think about it, already there are so many words, er... for example to do with computers, um... er... that are in English and that are used internationally, er... for example, um...'radio, television, football', these are all international words – English words though. So I think in years to come, um... there'll be very few people who don't speak English, not just a few words but, you know, whole sentences, even.
Q	And, er... do you think, Lynne, that teachers will start um... using English to teach other subjects, you know, for instance, geography or science, and that they'll be used in schools all over the world?
LYNNE	Yes, I do. I think that teachers will start experimenting with that. I think it is likely and I think in many ways it's the best way of learning English.

GREG Mmm...

Q Greg?

GREG Um... I'm not sure about that actually, I don't think that's right. I think some will be in English certainly, um... for example, lessons in science, say. But no, I think quite a lot won't be in English – other lessons. There's no reason why every single subject should be in English.

Q Right. Now, what about, um... British and American life and habits, institutions, do you think that it's important to know about those?

LYNNE I don't, not at all. I mean, I don't think that English as a language has anything to do with, you know, double-decker buses, and bowler hats, and hamburgers and yellow taxis. I mean, it's an international language and, um... it can be used for communication between, you know, people who don't know each other's language, um... as a tool really. So, I don't think that the cultural roots of English are important at all.

GREG Oh, sorry. Can I just come in there. I think that's... I really do disagree there, because I think you have to understand, er... the culture of a country, simply because there are some words that mean different things to different people depending on what country they're in, for example, er... the word 'tea', er... can be a drink to some people in one country and in another country it means an entire meal. Um... the word 'police' means different things to different people. You always have to know a little bit about the background and the culture of a country before you can fully understand the language.

Q Mmm. What about in the work, er... situation. How important is English there, what's its role?

LYNNE Well, I think it's really important and I think more and more people will use it at work – it's, er... easily understood wherever you come from and I think, well, actually, everyone will need to use more English for their work.

Q Mmm. Greg?

GREG Um... I think some people will need to use more English, particularly people working in big companies who have to travel a lot and do a lot of business between countries, but I think for the majority of the population in any country, um... who don't... who aren't involved in international business or moving around or travelling, then I think they'll be very happy sticking to their own language.

Q And the traditional language class as we know it – do you think that that will continue or will there be other forms of teaching, such as, you know, teaching involving television and computers, using those sort of technologies?

LYNNE Well, I think that the traditional language class will still exist. Er... I think that personal contact is very important with the language teacher and, er... of course, there is more than one person in a class, you can interact with the other students and I think that that's much more valuable often than just relating to a computer screen or, you know, listening to cassettes.

Q Mmm. Do you agree with that Greg?

GREG Not entirely. I think that we live in a computer age now and, um... it's highly likely that computers and other, er... videos for example – all those interactive programmes that you use with videos – will allow people to learn foreign languages in a different way on their own, um... so that you aren't dependent on teachers and other students. I'm not sure, but I think that's how it'll be.

Q And finally, can I ask you Lynne, do you think that, um... English will ever become more important than, um... the language of the native speaker.

LYNNE Well, no. I think obviously English is important, but I think your own language and your own culture and traditions are more important to you and I think it's good to respect those and to hold on to them.

GREG Yes, I agree. I think it would be very arrogant to think that English would be more important than your own language, I mean, 'cause your own culture and your personal identity and your national identity are, after all, far more important aren't they?

LYNNE Mmm. I think so.

Q Thank you very much.

GREG Thank you.

Lesson 12 Sounds, activity 1

First syllable actor banker chemistry secretary
Second syllable accountant arithmetic biology computer geography
Third syllable economics education engineer politician

Lesson 13 Listening, activity 3

CATHY So, what are you going to do this summer?

DUNCAN I'm going to visit South America.

CATHY Great!

DUNCAN I'm going to be travelling around quite a lot and I'll probably spend three weeks there.

CATHY So, where are you going?

DUNCAN To start off, I'm going to fly to Rio and perhaps I'll stay there for a couple of days.

CATHY Do you know anyone there?

DUNCAN No, but I've always wanted to go to Rio. It's meant to be quite spectacular. And I'll probably go up the Sugar Loaf mountain, with the cable car, like all the tourists.

CATHY And where are you going after that?

DUNCAN I'm going to fly to Santiago in Chile, where I've got some friends. We'll probably spend some time doing some sightseeing and then we're going to lie on the beach for a few days in Valparaiso, which is on the coast, not far away from Santiago. Then I'm flying to Lima where I'm going to meet my girlfriend and then we're going to visit Machu Picchu in the mountains.

CATHY And visit the ruins? Ah, fabulous.

DUNCAN Yes, it should be really interesting. And then we'll probably go somewhere on the Amazon, I don't know where yet, but I'd like to spend a week in the jungle. Then we'll probably fly home.

CATHY Well, have a nice time!

Lesson 14 Listening, activity 1

So, we're here in the very heart of Dublin now, with St Stephen's Green behind you, just over there. If you look to your left, the road over there is Dawson Street, and the street on the right-hand side from where we are now is Merrion Row. And if you look straight ahead, straight down this street, which is Kildare Street, you'll see the Irish Parliament building on your right, and at the end of the street, you can see Trinity College and College Park.

Lesson 14 Listening, activity 2

I think a good short tour of Dublin would start here in St Stephen's Green. Go along Merrion Row and turn left into Merrion Street and go straight ahead as far as Merrion Square. I suggest you walk all the way round Merrion Square, so turn right into Merrion Square South, left into Merrion Square East and left again into Merrion Square North. Go straight ahead into Clare Street and Nassau Street, and then have a walk round Trinity College and College Park. Then walk back along Grafton Street, and you'll find yourself back in St Stephen's Green.

Lesson 14 Sounds, activity 1

/ɑː/ article charm far bar
/æ/ bank map actor national
/ʌ/ pub Dublin number others bus

Lesson 15 Listening and speaking, activity 2

Q So, you live in The States. Where exactly do you live?

PAT In San Francisco.

Q Uh huh. An, um... we were wondering what, er... what typical meals were like in The States. I mean, what do you have for a typical breakfast?

PAT Well, a lot of people in The States eat a great deal, but I tend to eat a fairly small amount. Um... I'm quite attached to cereals still, and... so, I'll eat grape nuts or muesli, something like that. But, on the whole, I tend to just have, er... fruit and, um... juicing is very big in The States.

Q Uh, huh.

PAT I'll juice a lot of fruit in a blender.

Q Is there any special fruit that's really good?

PAT Um... apricots are very good. Apricots, orange, banana.

Q And, when you juice things, you make a mixture of these fruits, do you?

PAT Yes, you basically just bung everything in, and, um... juice it and then drink and then that keeps you going for most of the morning.

Q Lovely. How about lunch? Do you have lunch?

PAT Well, if I'm going to eat lunch, I tend to go more for, um... either light pasta dishes, or vegetable platter where you have, you know, beans, zucchini, egg plant, that sort of thing.

Q	Zucchini is... we call...
PAT	Zucchini is courgette.
Q	Courgette, and egg plant is...
PAT	That's aubergines.
Q	Aubergine, yeah.
PAT	I've obviously lived there too long, because I use those terms so naturally! But, um... more often than not, probably just a sandwich, which is usually like a French stick. I mean, they call them 'submarines' – they're big French sticks. Full of meat...
Q	And would that be your main meal?
PAT	No, no, no. The main meal would be in the evening.
Q	And what would you have then?
PAT	Um... well, San Francisco is... you can eat very well, um... if you go to seafood restaurants, so, and... I like a lot of seafood, so, um... perhaps a favourite is soft shell crabs.
Q	Lovely.
PAT	...with steamed vegetables. Perhaps have some sort of seafood chowder before that, and a mixed salad, and finish it all with key-lime pie, perhaps.
Q	Key-lime pie. What's that?
PAT	I've never quite worked out what it is. But it's got limes in it, and it's a bit like a cheesecake, with lime.
Q	Lovely.
Q	Karen, how long have you been living in Hong Kong?
KAREN	Just over a year now.
Q	And... and you've got used to the way of life there, have you?
KAREN	Yes, I think so. It's a very racy place, so you have to adapt.
Q	Yes, what... what sort of things do you eat? What's a typical breakfast for you?
KAREN	Um... well, during the week I'm very busy. I have to get off to, um... the school where I work quite early, so I have quite a quick breakfast – but a substantial one. Um... so usually cereal, toast, um... orange juice, a cup of tea.
Q	A bit like an English breakfast, really?
KAREN	Yes, yes. I live with two other English people, so we all eat an English breakfast.
Q	Yes, yes... And then, er... do you have lunch?
KAREN	If I have time, yes. I rush out and get a sandwich, or a baked potato from a local, um... fast-food place, if I can.
Q	Yes, yes. No Chinese delicacies?
KAREN	Not at lunchtime, no. No time!
Q	Yeah, that's not your main meal, then? The main meal would be dinner time?
KAREN	In the evening.
Q	Yes. yes. And then, what would you eat normally?
KAREN	Well, um... we get together – several teachers – and we go out to a Chinese restaurant, or um... there are plenty of other restaurants represented in Hong Kong, – um... Japanese, Indonesian – and we eat there. Um... a favourite Chinese meal would be 'dim sum', probably.
Q	Uh, huh. What's that?
KAREN	Well, it's like a dumpling, um... which they steam, and inside will be different types of meat or vegetables. It's very nice.
Q	Yes. And, are there any, um... deserts, any sweet things to eat?
KAREN	The Chinese aren't so good on their deserts, they're very sticky, and a bit too sweet for me, so I try to avoid those.
Q	Yeah. Thank you.

Progress check 11 – 15   **Sounds, activity 2**

/tʃ/ charm chicken cheese peach
/ʃ/ politician traditional she old-fashioned fish

Lesson 16   **Listening and speaking, activity 1**

Q	So, Ken, tell me. What is karaoke?
KEN	Well, basically, it's singing along to some recorded music. You have a microphone and there's some music playing and you sing the words – in tune, if you can.
Q	Where does this take place?
KEN	Well, all over Tokyo there are karaoke bars where you can go with friends, have a drink and sing karaoke. It's very common.
Q	Who performs it, then?
KEN	Anyone. Anyone who feels brave enough to sing in public, it could be you or me or anyone.

Q	It sounds as if you need a drink to sing in public. And what kind of music do you sing?
KEN	Well, it's traditional Japanese music for older people, but for young people it's mostly well-known Western pop songs, you know, Frank Sinatra, Phil Collins, Madonna, that sort of thing.
Q	Why do people enjoy it?
KEN	I don't know, really. It's a chance to show that you could be a pop singer as well, I suppose. It's also a way of showing how close you are to your friends. If you can make a fool of yourself in front of these people, then you really are good friends, that kind of thing.
Q	What about tango then, Philippa?
PHILIPPA	Tango is a very exotic kind of dance in Latin America, and it's especially popular in Argentina, where it came from originally.
Q	And where is it performed?
PHILIPPA	In concert halls or theatres, or maybe small bars.
Q	And who performs tango?
PHILIPPA	Well, in the theatre they're mostly professional dancers, although in dance halls and bars, everyone tries to dance the tango if the music is right.
Q	And what is the music they use to dance to?
PHILIPPA	Well, tango is both the dance and the music. You dance the tango to music specially written for it. They use the violin and the accordion quite a lot for it.
Q	And why do you think people enjoy it?
PHILIPPA	Well, it's a very passionate dance. It's full of life, it's great fun.
Q	Can you dance it?
PHILIPPA	Yes, well, not very well, but I try!

Lesson 19   **Sounds, activity 1**

/ə/ smaller bigger closer happier funnier
/ɪ/ smallest biggest closest happiest funniest

Lesson 19   **Sounds, activity 3**

1 He's less casual than she is.
   No, he's **more** casual than she is.
2 It's more noisy now than it was.
   No, it's **less** noisy now than it was.
3 Clothes are cheaper here than at home.
   No, clothes are **less** cheap here than at home.
4 He's less optimistic than she is.
   No, he's **more** optimistic than she is.
5 It's easier to get good clothes here.
   No, it's **less** easy to get good clothes here.
6 He's less confident than she is.
   No, he's **more** confident than she is.

Lesson 19   **Listening and speaking, activity 2**

Q	Oh, the one thing I don't like about Britain is the weather.
GRAHAM	Well, I agree. I've lived in a number of countries around the world, and I always like to come home to Britain, but I don't usually stay for long because I really enjoy the sunshine, and you just can't be sure of getting the sunshine at all in Britain.
Q	Even in the summer.
GRAHAM	Even in the summer. No, of course, sometimes we have a warm summer's day, but it can be really quite cold most of the time. I'd say the average temperature during the summer was about 18 or 19 degrees.
Q	Is it easy to get good clothes in Britain?
GRAHAM	I would say yes, it is quite easy, although obviously for really good clothes you have to pay quite a lot of money. But there's the tradition of the cloth industry, um... especially in the Midlands and the North, so at least good quality clothes are available, even if people don't wear them all the time.
Q	So you can get good clothes but you have to pay a lot for them?
GRAHAM	That's right, in proportion to what most people earn, anyway.
Q	Do people dress formally in Britain?
GRAHAM	In comparison with other countries, I'd say no, certainly not in social situations. I mean, it depends what you mean by formal, but apart from the office, most men don't wear suits, and many people come to dinner parties dressed very casually. Um... of course there are some

social circles where it's still very formal, but I'd say for the average Briton it's fairly informal.

Q With your experience of people from around the world, would you say the British are small or large people?

GRAHAM Er... to be quite honest, I think they are about average. Um... they're not particularly small, no.

Q And what is the design of clothes from Britain like?

GRAHAM I think it's quite good, especially for traditional clothes, er... Burberry, Jaeger and the kind of clothes people wear, or at least used to wear in the countryside. For town-wear and high fashion, there are now some very famous designers, whose clothes sell even in places like Paris and Milan. It's a small industry, but it's good quality, I think.

Lesson 20   Vocabulary and listening, activity 2

Three hundred and forty-six.
Six hundred and seventy-eight.
Two thousand, three hundred and forty-five.
Eighteen thousand, six hundred and sixty-four.
Twenty-four thousand, five hundred and eighty-nine.
One hundred and twenty-three thousand, four hundred and fifty-six.
Two hundred and two.
Fifty-four thousand, five hundred and sixty-six.
Three thousand, four hundred and eighty-one.
Four hundred and seven.
Ten thousand and twenty.

Lesson 20   Vocabulary and listening, activity 4

Q So, what was your most memorable journey, Sarah?

SARAH Well, I was a passenger on this incredible trip across America with a friend who was moving home from New Orleans to Laguna Beach on the California coast.

Q Mmm. How long did it take?

SARAH It took nine days in all. It was great fun!

Q What sort of distance is that?

SARAH About 2,500 miles, and that's pretty direct as well.

Q So where did you go?

SARAH Well, we set off from New Orleans and took the Interstate 10 Highway, which runs all the way from Florida to Los Angeles. We picked it up just outside the city, and we drove through Louisiana as far as San Antonio in Texas, where we stopped for the night.

Q How far was that, then?

SARAH Oh, the first day we did 550 miles.

Q What's San Antonio like?

SARAH It's quite interesting. It's a strange kind of mix of skyscrapers and Indian houses and churches. It's about 200 years old.

Q What's the scenery like in the country? Is it desert or mountainous or what?

SARAH Well, it's not really mountainous at that point, um... sort of hills and small trees. It's very remote, though, and you can drive hundreds of miles between gas stations, so you have to make sure you've got plenty of petrol.

Q It's cheaper there, isn't it?

SARAH Yes, the Americans call it 'gas', and it costs about a dollar a gallon, what's that – about 25 cents a litre, I suppose.

Q And, where did you go then?

SARAH Well, after that, we started to hit the desert proper and we drove 350 miles to Fort Stockton, which is a typical desert motel stop, in the middle of nowhere. Then we drove through the Guadeloupe Mountains National Park to El Paso, on the border between Texas and Mexico. Then between El Paso and Las Cruces you start climbing into the Sierra Madre.

Q Is it true that you can only drive at 55 miles an hour?

SARAH Well, that's the law, although a lot of people go a lot faster. It depends if you think there are any police patrols around.

Q Where to then?

SARAH We turned off and took a detour to Nogales on the Mexican border which was great. We had lunch there, and then headed north to Tucson in Arizona on Route 19.

Q And where did you stay en route?

SARAH In motels. They're incredibly cheap. The most expensive was thirty-five dollars, the cheapest was twenty. And from Tucson we turned west

again and crossed into California. From there it's only 300 miles to San Diego on the coast, which was very beautiful – my first sight of the Pacific Ocean. And then we drove on up the Pacific Coast Highway to Laguna Beach, not far from LA, and arrived at my friend's apartment, with a fabulous sea view and only ten minutes from the beach.

Q It sounds very memorable!

SARAH It was, it really was. But although it was great to arrive, it was much better to travel. Great fun!

Lesson 20   Sounds, activity 2

21	33	421	645	4,591
3,542	77,889	12,523	101,456	987,241

Progress check 16 – 20   Sounds, activity 1

/ʊ/ good book pullover took cook
/uː/ blue shoe boot suit

Lesson 22   Vocabulary and listening, activity 2

Q I expect you've seen many changes since 1989.

HEIDE 1989, that was a very important year for us. We were all very excited when the Wall came down of course, but we had no idea how different our lives were going to be from then on. Of course, the most important thing for my generation was the change of government and the fact that we've had an election for the first time in our lives. We now vote for our national representatives like every other country in the West, and of course we also elect our regional government as well. I was born ten years after they built the wall, in 1971, so I have never lived in a democratic country until now.

Q Have the changes all been good ones?

HEIDE Well, no, I don't think so. We had a fairly good standard of living under the old government, and although we have more possibilities now, the fact is that we haven't got the money to take advantage of them. And then unemployment was unthinkable in the old days, but now it's very common. My father, who was a teacher, lost his job. Just imagine a teacher losing his job! And, there's been a kind of inflation as well, with prices going up so that they match prices on the western part of the country. It's quite hard at times.

Q Do you get the chance to meet people from other countries and talk about standards of living and the lifestyle in their countries?

HEIDE Yes, we do. We can travel now of course, although as usual, money is a problem. And we have a lot of tourists who come to visit us now, which never happened before. So, we get the chance to talk to people from other countries.

Lesson 22   Vocabulary and listening, activity 3

Q Has anything happened to you personally in recent months?

HEIDE Well, for me personally, the most important thing is that I've got married and I've just had a baby boy, called Konrad, who is now one month old. He's totally changed my life, but it's worth it. I mean, my life is very different now, because before I got married I lived with my parents in a very small flat in the east of the city, But now, of course, I live with my husband and Konrad and we've just moved to a new flat in Potsdam, in a new building. It's very comfortable for the three of us. What else? Oh, I've found a new job – it's for a computer firm and I start on Monday. I've bought a new car – a Volkswagen. And I've given up smoking. My husband doesn't smoke and I had to look after myself when I was expecting Konrad, so I've given up and I feel much healthier now.

Lesson 23   Vocabulary and listening, activity 3

Q Are there any, um... special events or festivals which everyone in Australia celebrates?

BARRY Well, there's Christmas and Easter, like you have in Britain, I mean. They're pretty important.

Q I was thinking more of national holidays, like Independence Day or events that happen in your town or region, you know, local festivals or something like that. I don't even know if you have an Independence Day in Australia.

BARRY Well, that's because we're not really independent, are we? At least, not yet. But I suppose there's Australia Day, which we celebrate on, er... the twenty-sixth of January.

Q Oh, and what does that celebrate?

BARRY It's the day Captain Cook arrived in Botany Bay in Sydney in 1788, bringing Europeans to Australia. But the trouble is, we don't really celebrate it very much usually, except on special anniversaries, er... like in, er... 1988 which was two hundred years since he arrived. Um... no, there's one day which people really enjoy and that's the Melbourne Cup in November.

Q Oh, and what does the Melbourne Cup celebrate?

BARRY It's a horse race.

Q A horse race?

BARRY Yeah, the whole country stops during the horse race, and everyone wants to know which horse wins. We love horse racing.

Q And, er... you say it takes place in November?

BARRY Yeah, the first, um... the first Tuesday in November. At two-forty in the afternoon, every year.

Q And, er... when did it first take place?

BARRY I think it started in 1874.

Q Right, so, it's over a hundred and twenty years old. And what exactly happens?

BARRY Well, people from all over Australia come to Melbourne on special planes and trains for the day and dress up and go to Flemington race course. And they take picnics in their cars which they eat before the race. Um... and everyone bets on the horse they think is going to win. And then the race starts.

Q At exactly two-forty?

BARRY That's right. And at two-forty-three, it's all over.

Q So, it only takes three minutes?

BARRY Yes, and if you're enjoying yourself too much, you miss it. It's great fun. It's a great social occasion, a kind of, er... social ritual at the start of summer.

Q Is it a public holiday?

BARRY It is in the city of Melbourne and the whole of the State of Victoria. Everybody takes the day off. But not in the rest of Australia.

Q But, so, even people who can't go to Melbourne are interested in the race?

BARRY Oh yeah. The interesting thing is that the whole of Australia wants to know who wins the Melbourne Cup. Everybody listens to the race on the radio or watches it on television. The traffic stops and in Canberra, the politicians stop work in parliament.

Q So everyone's involved, even people outside Melbourne.

BARRY That's right. It's a kind of state occasion for the whole of Australia.

Lesson 24   Vocabulary and listening, activity 2

WOMAN 'Bill'. A 'bill' is paper money. You have a 'dollar bill', a 'five-dollar bill' and so on.

MAN Right... we call that a 'bank note'. Um... 'trousers' are... an item of clothing. Yes...

WOMAN Oh, I know what trousers are! Yes... we call them 'pants'.

MAN Oh, right.

WOMAN Oh, the snack food that's round and flat and fried and thin and very crisp, we call them, er... 'chips... potato chips'.

MAN Oh, right. Er... cold, you mean? Yeah, we call those, um... 'crisps'. You buy them in a packet... 'crisps'. Um... er... 'car park', well... a place where you park cars.

WOMAN Yeah, we call that a 'parking lot'. Um... if I need some medicine or something like that, I go to see the 'druggist'.

MAN Right. Oh, is that the person, or... is that a place or a person?

WOMAN No, the place is the 'drugstore', the 'druggist' is the person who'll give me the medicine.

MAN Right, we call that a 'chemist' – but that's the name of the shop. Um... a 'state school' is a school which is, um... funded by the state, it's the opposite of a private school, in other words.

WOMAN Oh yes, we call that 'public school'.

MAN Oh, right.

WOMAN Um... water in a sink comes out of a 'faucet'.

MAN Ah yes, we call that a 'tap'. Um... 'traffic lights' – do you know what those are? When you're driving along the road, and you have to stop because there are lights...

WOMAN You have to stop, so we call them 'stop lights'.

MAN 'Stop lights'. Right...

WOMAN Um... when I get a hamburger I also like to get 'french fries', which are the strips of, er... fried potato.

MAN Oh, right – 'chips', we call those.

WOMAN Do you?

MAN Yes, 'chips'. Um... when you travel around, for example in London on, er... on the train under the ground, that's called the 'underground'.

WOMAN No...

MAN Yeah.

WOMAN It's called a 'subway'.

MAN No. The 'underground', yeah...

WOMAN Well, that's what we call it. I fill my car with 'gas'.

MAN Ah yes, we call that 'petrol'. Um... there's another item of clothing – er... a 'waistcoat', um...

WOMAN Oh, yeah, that men wear. We call that a 'vest.'

MAN That's right, it doesn't have any sleeves, yeah? 'Vest'... yeah.

WOMAN Every town in America has a 'main street' where all the shops and things are.

MAN Oh, right... No, we call that the 'high street'. Same thing, 'high street'.

Lesson 25   Listening, activity 2

SPEAKER 1 Well, it's... it's that stuff you need to put two different pieces together. For instance, two pieces of paper. You put that stuff on one bit of paper and stick the other paper on top of it, for example. Or, you can do that with leather as well, if your shoe gets broken, or you can do that with wood, and things like that.

SPEAKER 2 It's a piece of material. Um... it's a square and it's soft and you use it to... after a bath for drying yourself when you are wet.

SPEAKER 3 It looks like little pieces of wood, very thin little pieces all in a box and, er... at the tip there's a... they are either black or red and it's something you use to light a fire or anything like that.

SPEAKER 4 Er... I want... you put it on when it's hot and you buy it in a bottle, a plastic bottle and you put it on your body and it protects you from the sun.

SPEAKER 5 It's a machine for cleaning. You have a tube and it, er... it sucks the dust. It's a machine for cleaning the carpet or the floor.

Progress check 21 – 25   Sounds, activity 1

/ɜː/ surfing fireworks learnt heard university word
/ə/   primary economics lecture chosen

Progress check 21 – 25   Sounds, activity 2

/ɔː/ sport awful walking performance four
/ʌ/   up summer bus pub sunny

Lesson 26   Vocabulary and listening, activity 3

Conversation 1

POLICE OFFICER Good morning, sir.

DRIVER Good morning, officer.

POLICE OFFICER Do you realise what speed you were driving at, sir?

DRIVER Er... about a hundred and ten, I suppose?

POLICE OFFICER No, sir. When we were following you in the fast lane a few minutes ago, you were driving at one hundred and eighty.

DRIVER Was I really? Oh, goodness me! Dear, oh dear! I didn't know this car could go that fast! Aren't these German cars magnificent!

POLICE OFFICER Yes, sir. But we're not on a German autobahn now. We're on a British motorway. You really mustn't drive so fast. You mustn't drive at speeds over one hundred and ten in this country.

DRIVER Ah. Er... my wife's ill.

PASSENGER No, I'm not.

DRIVER We're going to the hospital and the traffic delayed us, so I was driving fast. What did you say, dear?

PASSENGER I'm not ill.

DRIVER Ah.

POLICE OFFICER And did I see you using your car telephone?

DRIVER I was phoning the hospital.

PASSENGER No, you weren't.

POLICE OFFICER	You have to have both hands on the wheel when you're driving. You mustn't use the car telephone, you know. Can I see your licence, sir?
DRIVER	My brother's a policeman.
POLICE OFFICER	Is he sir? How interesting.
PASSENGER	No, he isn't.
DRIVER	Aargh! Will you be quiet!
POLICE OFFICER	I think you'd better follow me, sir...

Conversation 2

POLICE OFFICER	Oi! You! Stop!
CYCLIST	Who me?
POLICE OFFICER	Yes, you. Did you see that red light?
CYCLIST	What red light?
POLICE OFFICER	The red traffic light. Even cyclists have to stop at red traffic lights. And have you got any lights?
CYCLIST	No.
POLICE OFFICER	Well, you have to have lights when you ride a bicycle at night.
CYCLIST	Is that so?
POLICE OFFICER	And I saw you riding on the pavement a moment ago.
CYCLIST	Did you?
POLICE OFFICER	You mustn't ride on the pavement.
CYCLIST	Officer, do you know who I am?
POLICE OFFICER	No, madam, but I don't care if you're the Queen of England. Could I have your name, please?
CYCLIST	No, you can't. You'll have to catch me first.
POLICE OFFICER	Oi! Come back here!

Conversation 3

TICKET INSPECTOR	Tickets, please. Tickets, please. Good morning, can I have your tickets please?
PASSENGER	Here you are.
TICKET INSPECTOR	Excuse me, but this is a second class ticket.
PASSENGER	I know.
TICKET INSPECTOR	But you mustn't sit in here. This is a first class carriage.
PASSENGER	But there weren't any seats in the second class carriage. It's standing room only.
TICKET INSPECTOR	I'm sorry, madam, but you either have to go back to the second class carriage and stand, or pay for a first class ticket.
PASSENGER	But I've paid for my ticket. You have to give me somewhere to sit.
TICKET INSPECTOR	I'm sorry, but in Britain the law is your ticket is for transport by train only. It does not give you the right to a seat as well.
PASSENGER	Oh, but that's outrageous!
TICKET INSPECTOR	I'm sorry, but I don't make the rules.

Lesson 27   Listening and speaking, activity 2

The Skylight, part 1

The heat, as the taxi drove up the hill, became more violent. The woman sat in the back of the car with a five-year-old boy beside her, his thumb in his mouth.

'When are we there?' the boy asked.

'Soon.'

The child's eyes closed. 'Oh no', she thought. 'He mustn't go to sleep'. She could hear herself telling the story in the cold English spring. 'It's so sensible to take this house for the summer. It's in the mountains, ten minutes drive from Golfe-Juan. Philip will drive the girls but I will take Johnny by air. And the Gachets will have everything ready for us.' But now it was real. She was hot and afraid. Will everything be all right when we get there, she wondered. Suddenly, the driver turned off the road, drove up a narrow track, and stopped. The woman could only see stones and grass.

'But – where?'

He pointed and got out. He picked up their suitcases and walked away. She took the child's hand and followed the driver. Above them, on the terrace, was the square grey house. A small skylight in the roof caught the sun. The shutters and doors were all closed.

'*Vous avez la clef, madame?*'

'The key? But Monsieur and Madame Gachet are expecting us.'

The driver tried the door. It was locked. She knocked. There was no answer.

Lesson 27   Listening and speaking, activity 4

The Skylight, part 2

The driver wanted his money. She paid him and he disappeared. She heard the taxi leaving.

'Why can't we go into the house?' asked Johnny.

'Because it's locked.'

She looked up and saw the skylight.

'If there was a ladder, perhaps we could...'

'There's a ladder.' he said. 'Can we lift it?'

It was quite light.

'Are you going to climb up there?' the child asked.

She hesitated. 'Yes, I suppose so.'

She started to climb. At the top she saw the tiny skylight was open. She couldn't get through it, but a child could do it. She could lower Johnny through, and he could run downstairs and unlock a window. She came down the ladder. He was lying on the ground, nearly asleep.

'Johnny,' she said, 'Would you like to climb the ladder?'

'Can I climb it now?' he asked.

'Yes. Yes, you can. When you've got through the skylight, I want you to do something.' She explained, very carefully.

Together they climbed the ladder. She lowered him through the window until he stood on a table.

'Can you get down?' she asked.

'Yes. Shall I go and open the window now?'

'Yes,' she said, 'And hurry.'

She climbed down the ladder, went to the window and waited. It was getting dark.

'Johnny, it's this one. Are you there, Johnny?'

Give him time, she thought. He's only five. He can't hurry. She climbed up the ladder again and shouted, 'Johnny, can you hear me?' Her voice had no volume, no echo.

Lesson 27   Listening and speaking, activity 6

The Skylight, part 3

It was now dark. She went down again and ran round the house, shouting his name. Something has happened to him, I must go for help. She ran to the road and when she saw the lights of the car, she waved her arms to stop it. She started to cry. It was a long time before the three men understood.

'But how can we get in? We have no tools,' they said.

'There's a farm back there. Will you take me?' They let her into the car.

'Turn round. It's back there on the left. There it is!'

They turned off the road. She got out of the car and ran to the front door. A small woman in trousers opened the door.

'My dear, what's happened?'

'You're English?' She told her the story. Another woman appeared.

'Yvonne,' Miss Jardine said, 'Get some tools, a hammer and an axe.'

They all got into the car and went back to the house. They drove up the lane and stopped. She ran to the house, calling 'Johnny? Johnny?'

Lesson 27   Listening and speaking, activity 8

The Skylight, part 4

One of the men took an axe and smashed the shutters. She was quickly through the window.

'Johnny, where are you?'

She ran up the stairs. A door on the first floor was open. He was lying on the floor, fast asleep. Surrounding him were lots of toys. She shook him gently. He opened his eyes.

'I like the toys,' he said.

His thumb went back into his mouth and his eyes closed again. She sat with her head on her knees, her arms round her body.

'Oh, thank God, She whispered. 'Oh, thank God.'

Lesson 28   Listening, activity 1

| Q | I mean, in Britain for instance, if I, sort of... want to cross the street at a pedestrian crossing and, er... the light is red but there are no cars, is it allowed to cross? I mean... |
| JANE | Oh yes... yeah, you don't have to wait for the pedestrian light to turn green before you cross, like you do in some countries. |

Tapescripts

Q Like Japan you do, and I think Germany you do as well.

JANE Yes. But no, you can cross.

Q I see... um... and if... when I'm in a park, for instance, you know, with grass everywhere, um... can I walk on that grass?

JANE Yes, unless there's a sign that says 'Do not walk on the grass', usually you can always walk on the grass.

Q Ah, because in France it's virtually impossible...

JANE Really?

Q I mean, you'd get a fine, except a few areas. But, that's good... Now, I had a problem last time I wanted drink, it was, um... I think, Sunday in the afternoon, er... around half-past three, and, er... I mean... was it possible to get a glass of beer somewhere because I found it very difficult?

JANE Ah well, not on a Sunday, probably on any other day it would be OK, I'm talking about pubs.

Q Yes, because everything was shut.

JANE Exactly, but there is a possibility, um... of getting a beer or an alcoholic drink in a restaurant.

Q But I'd have to eat then.

JANE But you usually have to eat something, that's right. That's usually the rule.

Q What a bore! Mmm. Yes... And, I wanted to ... I mean, I was told that if you had young children you can drink in a bar? Er... I mean, can you take the children with you, or...?

JANE Mmm... not into the bar. Sometimes they might have a beer garden or something like that. But if you're talking about very young children – no, they can't come in with you, not to a bar.

Q Oh dear, so they have to go to separate place if there is such a thing then?

JANE Well, there isn't usually.

Q There isn't!

JANE There isn't usually, no.

Q And, er... in a restaurant, er... perhaps foreign food or whatever, if I don't know the food, can I go into the kitchen and look at the food?

JANE And have a look?

Q Yeah, have a look, I mean, sometimes in France we do that... it's allowed – not every restaurant – is it possible here?

JANE I've never seen it done.

Q If you ask nicely?

JANE I think if you asked very nicely, um... they might say yes but I think normally I would say no.

Q You just wait for a plate? Oh, I see. And, for all my correspondence and letters and postcards, um... is it possible in England to go to a newsagent or a tobacconist shop and get some stamps there?

JANE To get the stamps? Yes, yes, um... also, all kinds of places now – supermarkets sell them. Yes, there's usually a sign on the door that indicates that they will sell them.

Q A friend of mine came here about ten years ago, and I think it wasn't the case then.

JANE That's right.

Q So you've changed!

JANE Yeah, we've changed!

Q Yes, and what about smoking in the cinema, I mean, if there's no sign for instance which says 'no smoking' is that allowed, you can actually smoke if there's no sign saying you can't?

JANE Well, yes I think so. Usually I think most of the cinemas are non-smoking now, but then, yes, then there would be a sign. So if there's no sign...

Q You can get away with it?

JANE I suppose you can smoke and hope nobody says anything. Yeah. Yeah... I think you can.

Q And, um... in a cafe for instance, if I want a drink and I don't want to wait at a table or something, can I just go to the bar and, er... pay for it there?

JANE Oh, I think you... Yes.

Q That's no problem? It's the same.

JANE Yes, I think you can do what you want in most cafes.

Q And, for instance, If I feel a bit ill, er... and I haven't got medical insurance, or much money on me, or... er... Can I make an appointment to see a doctor, any doctor?

JANE A doctor? Oh yes, Yes, you can, er... get an appointment at any doctor's surgery.

Q No problem?

JANE As a temporary patient, it shouldn't be a problem. It is possible.

Q Oh, that's wonderful. Oh, that's good. And lastly, if I want to take a bus, can I buy a single ticket before I get on? Is that possible?

JANE A single ticket? Er... I don't think you can on the bus.

Q So I have to buy it on the bus itself?

JANE Yes. You can get passes that allow you to go on the bus all day, but just for one single ticket, you must buy it as you get on the bus.

Q I see. Mmm. Thank you very much.

JANE OK.

Lesson 29 **Listening, activity 2**

Q So, Doctor Samuels, what do you advise for people who want to avoid jet lag?

DOCTOR The thing about flying is that it has a dehydrating effect on the body, and this means that you need to replace the missing liquid with lots of water. It's also better to avoid alcohol, as this only makes the dehydrating effect worse. So, lots of water or juice, and adjust your body clock by adopting the time of your destination as soon as possible. So even if your body tells you it's time for bed, try and stay awake until it's bedtime in your new time zone.

Q And what can you do to avoid stomach upsets?

DOCTOR Mmm. One of the main causes of stomach upsets when you're in a foreign country is the water supply. So, the most important thing to do is to drink only bottled or boiled water, and don't forget the ice cubes in your drink too – they may not be from a clean source of water. And don't eat uncooked food, like salad, because it may have been washed in dirty water.

Q I always get badly bitten when I'm away. What do you suggest for this?

DOCTOR Well, keep your arms and legs covered in the evening when mosquitoes like to bite most of all, and maybe wear a hat as well. Insect repellent is very useful but you'll find the most determined mosquito will always find a patch of skin to bite.

Q And sunstroke?

DOCTOR Oh, well, obviously the most important thing is not to spend all day in the sun. If you come from a country where you don't get much sunshine, I suggest you spend only about twenty minutes or half an hour in the sun during the first day or two and gradually increase the time. And you should wear a hat, because your head is where you're most likely to catch the sun.

Lesson 30 **Reading and listening, activity 2**

WOMAN OK, here we go, um... question number one: 'You're a guest in someone's home, you'd like a cigarette, what do you say?'

MAN Um... I think it's got to be 'a', don't you? 'Is it all right if I smoke?'.

WOMAN I think that's most polite, yes. 'Is it all right if I smoke?' I'll tick that.

MAN OK, number two: 'A friend suggests you have dinner together in a certain restaurant, at the end of the meal the waiter brings the check, what do you say?'

WOMAN Ooh, um... well...

MAN Well, it could be 'c', because he did offer...

WOMAN 'Your friend suggested dinner, and you expect him to pay.'

MAN ...but I think 'b' is...

WOMAN 'Shall we share this?'

MAN It's... yes. Yeah. 'Shall we share this?' He could always persuade you later, couldn't he?

WOMAN Question number three: 'You're visiting a friend when the phone rings, what do you expect her to say to the caller?' Oh, 'b'.

MAN I think it has to be 'b'.

WOMAN It has to be 'b'.

MAN Everything else is very bizarre.

WOMAN 'Would you mind if I called you back, I've got a visitor here at the moment.' Yes.

MAN Uh... number four: 'It's late and your neighbours are playing very loud music, what do you say to them?'

WOMAN 'Turn down the music!'

MAN Yes! No, they're your neighbours, you have to try and get along. I think you'd start with 'b'.

WOMAN 'Could you turn the music down please?' Number five: 'You meet a Ms Esther Craig for the first time, you don't know how to address her, what do you say?'

MAN	This one's a little odd, isn't it? 'What do I call you?'
WOMAN	No.
MAN	'Would you mind telling me what to call you?'
WOMAN	No.
MAN	Do you think it's...
WOMAN	I think it's 'b'.
MAN	'b'? 'Can I call you Esther? ... Can I call you Esther?' It's friendly. It's...
WOMAN	Yeah, 'b'.
MAN	Yeah. Six: 'You meet someone at a party and get on very well, as she leaves she says, "Nice meeting you, we must do lunch sometime."'.
WOMAN	Do lunch?
MAN	Do lunch!
WOMAN	Ooh, um... I like 'c'. 'That's a great idea, bye.'
MAN	But, that sounds like a kiss-off because you don't... she doesn't... you don't know their phone number. It's got to be 'b' if you're serious about it.
WOMAN	'b'. 'Would you mind giving me your phone number?' You're right, 'b'. Number seven: 'You'd like your friend to lend you a book, what do you say?' Oh, 'c'.
MAN	Well 'c' seems the most polite – 'Would you mind lending it to me?'
WOMAN	Yes, 'c'.
MAN	Very direct. Number eight: 'You're at the information office at the railway station and you want to know some train times. What do you say?'
WOMAN	Mmm.
MAN	Well, 'c' seems a little formal, and English!
WOMAN	'I wonder if you could tell me when the next train to New York is?' Yes, I'd say 'a'. 'When's the next train to New York?'
MAN	'a'. When's the next train to New York?' As long as you don't say it too rudely! Say it nicely, 'When's the next train to New York?'
WOMAN	Question number nine: 'Your host serves you food you don't like (ugh!), you eat it but then the host offers you more, what do you say?'
MAN	It's got to be 'a'.
WOMAN	It's got to be 'a'.
MAN	'It was very nice, but no thank you, I've had enough.'
WOMAN	Yes.
MAN	I don't think you really need to have eaten all of the first course, even. Uh, number 10: 'As you are leaving a shop the assistant says, "Have a nice day." What do you say?'
WOMAN	I like 'c'. 'No thanks I've made other arrangements.'
MAN	Yes, but that might be a little too ironic for the shop assistant, so maybe 'a', I think?
WOMAN	'Thank you, same to you, bye.' Yeah.
MAN	Nice and direct, simple, polite.
WOMAN	I think we passed.
MAN	I think so.

Progress check 26 – 30 **Sounds, activity 1**

/əʊ/ wrote vote know only telephone home photo
/ɔɪ/ boy noise royal unemployment

Progress check 26 – 30 **Sounds, activity 2**

/ʃ/ shoe station pressure situation
/tʃ/ teacher temperature
/dʒ/ oxygen passengers stranger

Lesson 31 **Listening and speaking, activity 3**

Part 1

I was at home one afternoon sitting in front of the fire and watching the television, when suddenly there was a knock at the door. I wasn't expecting anyone, so I was quite surprised. I went to the door and opened it, and there was the Queen of England standing at the door, wearing her crown and holding a Harrods bag, as if she was coming home from the shops, just like you or me.

'Hello, Queen,' I said. 'Do come in.'

'Thank you,' she said, and came in.

I showed her into our front room, which we only use when we've got company – after all, it's not every day you have the Queen of England come to visit you, is it?

'This is very good of you,' she said in that voice of hers – well, she

couldn't say it anyone else's voice, could she?

'I went shopping this afternoon, and I am so tired that I really must sit down and rest my feet for a moment.'

So I said to her, 'You sit down, Queen and put your feet up. Would you like a nice cup of tea?' And, er... well, she smiled in that lovely royal way she's got and said, 'Oh, thank you.'

I went into the kitchen and got out the special china tea cups and saucers, it's called 'Royal Doulton', just to make her feel at home, and some chocolate biscuits. I was getting the tea ready when she called out to me, 'The shops are so crowded at this time of year.' So I went back into the front room.

'Yes, they are. What have you bought?' I asked, looking at her Harrods bag.

'I was looking for some curtains when I saw this lovely material, and I thought to myself, that would look very nice in St George's Hall.'

So I asked her, 'Where's that?'

'Windsor Castle,' she said. 'We were planning to redecorate the castle when the fire burnt the place down. Bit of luck, really.'

I said, um... 'Ooh, yes, it gives you a chance to start again, doesn't it?'

Lesson 31 **Listening and speaking, activity 5**

Part 2

I went and got the tea and brought it back and poured her a nice cup of tea and gave it to her, and, er... we carried on talking just as if we were old friends, which, in a way, I suppose we are. But, er... when you think about it, it's quite amazing, isn't it? I mean, there was the Queen sitting on our sofa in our front room. Anyway, we were talking about the weather when suddenly the phone rang. I picked it up. It was Prince Philip.

'Is the Queen there?'

So I said, 'Yes, she is' and gave her the telephone.

'Hello, Philip, I won't be long. Just having a cup of tea. I'll be back at about five o'clock if I don't wait too long for a bus.'

She came back to the sofa and was just finishing her tea when my husband arrived home from work.

'This is the Queen, dear,' I said helpfully. 'She's just been shopping and now she's going home.' So my husband said, um...

'Oh, hello, Queen, pleased to meet you,' and shook hands. 'Would you like a lift home?'

Now my husband never gets the car out in the week, so I knew that this was a very special occasion.

'Thank you so much, but I'll get the bus. The 75 is usually quite good at this time of day,' she replied.

And, er... well, then, she stood up and said thank you and goodbye. Oh, and she was going down the garden path when she stopped, turned to me and waved to me like she does on television. And she was gone!

Lesson 35 **Listening and speaking, activity 1**

Q	Now, Stephen, I'd really like to know, what do you say at the start of a meal in England?
STEPHEN	Uh, you don't really say anything, actually. I mean, you can say, 'Oh, this looks delicious', or something like that, but there's nothing formal that you say.
Q	Really? Because in Germany it's 'guten appetit', so is there such a thing as 'good appetite'?
STEPHEN	No, no. Nothing like that.
Q	Oh, that's surprising. Mmm. And, um... what time, roughly, do you have lunch and dinner?
STEPHEN	Um... I'd say that you have lunch at round about one o'clock, um... and dinner at about seven o'clock. I mean, obviously sometimes it can be later, eight o'clock, even nine o'clock, but normally I would say about seven o'clock.
Q	Right, well that's roughly the same in my country, I think. And, um... tell me, how long does a typical lunch or dinner last?
STEPHEN	I would say that lunch and dinner, I would say they last about half-an-hour, thirty minutes...
Q	Is that all?
STEPHEN	Yes, if it's just an informal lunch. I mean, obviously if it's a dinner party or a special occasion, um... it would last longer. But if it's just a...
Q	Like a family meal?
STEPHEN	Yes, but if it's just an everyday family meal, then half an hour.
Q	Ah, right. And, um... now that's just my curiosity, in which hand do you hold your fork?

STEPHEN   Ah, that's simple, you hold your fork in your left hand and knife in the right hand, always.

Q   Yeah, that's the same in my country, yeah. And, um... tell me, Stephen, do you actually use a napkin, like, generally in Britain? And if you do, tell me, where would you put it?

STEPHEN   Uh, some people do, and some people don't – there are no rules really. Um... if you're in a slightly posh or smart place you're probably more likely to have a napkin. If you do use a napkin you put it on your lap.

Q   Right, um... so you wouldn't put it round your neck, like the French?

STEPHEN   No, well, generally not.

Q   Rather on the lap. And um... tell me, at which meal would you eat the following food, um... melon – when would you eat that?

STEPHEN   Melon, well, you could really eat melon for breakfast, um... or for lunch or for dinner, any of those three.

Q   Oh, right. And how about pasta?

STEPHEN   Pasta, I would say lunch certainly and dinner, but not breakfast.

Q   Right. And how about fish?

STEPHEN   Fish, you can have fish for breakfast or lunch or dinner. Yes, all three.

Q   Oh right. And um... how about steak?

STEPHEN   Steak, er... lunch or dinner, not breakfast, though apparently in America they do, but not in this country.

Q   All right. And, um... tell me, where do you actually put your knife and fork once you have finished your meal? Is there a specific way you would put them in England?

STEPHEN   Um... yes, there is really. What most people do is they put the knife and fork together in the middle of the plate so the handle's pointing towards them and the points are facing away from them.

Q   Oh, is that right? Because in Germany you couldn't do that, you'd have to put it slightly sideways; both of them parallel but slightly sideways, it couldn't be in the middle. And, um... yeah, and how about your hands? I mean, during a meal, I mean, or... where would you put your hands when you're at the table but not in the process of eating?

STEPHEN   Um... I think most people would put them just on their lap.

Q   Really, now, that's interesting because, again in Germany you couldn't do that. You'd have to have your hands on the table, sort of just loosely just lying on the table; they couldn't be under the table.

STEPHEN   No, I think most people'd have it most certainly underneath, just on the lap.

Q   Uh huh. And um... tell me, how do you eat cake? Do you eat cake with a fork or a spoon?

STEPHEN   Um... I think informally you eat cake with your hands, with your fingers. Um... in a slightly more formal situation or if the cake is particularly sticky or messy you might use a fork. Um... so it's either a fork or hands really, a spoon you normally eat things like pies or puddings.

Q   Right. Would it be a smaller fork like a cake fork or would it be normal size?

STEPHEN   Yes, it would be quite a small fork.

Q   Yes, that's the same, actually the same, in Germany. And, um... because you were just saying you eat cake, you can eat cake with your hands. What food do you usually eat with your fingers at the dining table?

STEPHEN   Um... chicken, chicken bones normally, you know, bones with the chicken still on them. Um... bread, bread and butter, sandwiches, that kind of thing. Cheese, pieces of cheese. Er... fruit certainly, um... cake, yes, those are the main ones I'd say.

Q   Yeah, it's funny. Do you have that... in Germany it would be even rude to eat chicken with a knife and fork.

STEPHEN   Oh, really?

Q   You have to, I mean, you can eat the chicken breast, of course, but the legs and the wings you would have to eat.

STEPHEN   You'd almost have to pick it up?

Q   Yes, you have to.

STEPHEN   Not quite the same here.

Q   Right. And, um... tell me, what... are there any times of the day when you usually drink coffee and tea?

STEPHEN   Um... well, anytime really. Um... people drink coffee at anytime certainly, breakfast, mid-morning, after lunch, in the evening. Um... tea – people... a lot of people drink tea at teatime obviously, and some people have tea for breakfast.

Q   Right, so that would be a little more specific than the coffee?

STEPHEN   A little bit more, yes.

Q   And tell me, when can you actually smoke during a meal?

STEPHEN   Ah, well, you can't smoke during a meal really, or you don't. People smoke before a meal and after, but not during it.

Q   And if that meal consists of several different courses and if it lasts, like, for a few hours?

STEPHEN   Um... no. Sometimes if there are lots of people then you might have a break between courses when people can smoke, but not always.

Q   Ah, right. Because in Germany you could do that. Between courses that would be no problem.

STEPHEN   Yes, well that's acceptable here too.

Q   And, um... tell me, what do you actually say... what is the word you use when someone raises their glass?

STEPHEN   Oh, you mean to toast someone? 'Cheers!'. 'Cheers' is what you say.

Q   And, um... tell me, do you actually, do you have soup in the summer?

STEPHEN   You can do. Soup is normally drunk in the winter but quite a lot of people have soup in the summer. You can have cold soup, of course, like 'gazpacho' or something like that, but people occasionally drink warm soup as well. It's perfectly possible.

Q   Right, so that's an all-year-round thing really, and... how about salad, um... would you eat salad in the winter?

STEPHEN   Sometimes, yes. It's like soup really, um... salad is generally eaten in the summer, most often but, er... you can certainly have it in the winter.

Q   Right, well, thank you very much Stephen. That was interesting.

Progress check 31 – 35   **Sounds, activity 1**

wait wallet move visit want work sandwich drive invite women wear shiver walk behave

Progress check 31 – 35   **Sounds, activity 2**

1 ear   2 hair   3 eye   4 hat   5 hate   6 eat   7 art   8 as

Lesson 37   **Listening and speaking, activity 1**

Part 1

I was on holiday in the Lake District, and I was visiting a church in Kendal. I was sitting in the churchyard, relaxing for a moment, and my handbag was beside me, although I was holding the strap. Not the sort of place where you expect anything to happen, is it? Suddenly, someone came up from behind, grabbed my bag and pulled it very hard, breaking the strap. I shouted, first in pain, because when he pulled the bag it hurt my wrist, then in anger as I saw him get on a motorcycle and drive away. I felt awful as I watched my passport, my money, credit cards, various documents disappear down the road. The police were very kind and said that this sort of thing happens too often these days. I thought to myself, 'If I ever catch him, I'll kill him!' I told the consulate about the loss of my passport, and I cancelled my credit cards, got some more money and tried to forget about it. But that wasn't the end of the story.

Lesson 37   **Listening and speaking, activity 3**

Part 2

Four days later, the police rang me at my hotel and said they'd got some good news. A young man was trying to change some Australian money at the bank in Windermere. Now, there aren't many Australians in the Lake District at this time of year, and secondly, the young man wasn't Australian. So the bank clerk called the police, who came very quickly and they stopped the man as he was walking away from the bank. When they questioned him, he broke down and admitted he was guilty. They asked me to go to the police station in Kendal to identify him. Well, when he took my bag, I didn't see his face, so I couldn't really say if it was him. But he recognised me, and said, 'I'm sorry, I'm really sorry.' The police showed me the other things from my bag which he had on him. The man started to cry. The police said he came from Manchester. He was unemployed and he had a family to look after. I was the victim all right, but now it was me who was feeling sorry.

Lesson 37　Listening and speaking, activity 5

Part 3

The police said 'If we let him go, he'll probably take someone else's bag in some other town. But if we send him to court, he'll get a fine, which he won't be able to pay, so he'll go to prison. If he goes to prison, either he'll never take anyone else's bag again, or he'll learn how to do it more efficiently. So what do we do?' I didn't know what to say, so... I just felt so guilty and I had to keep telling myself, 'He's the criminal, I'm the victim.' Well, in the end, they sent him to court, and he got a fine, which was small enough for him to pay, but now he's got a criminal record, and will probably try to take someone's bag again. It's crazy. He's sorry. I'm sorry. We're all sorry.

Lesson 38　Listening, activity 1

ALEX	I love coconut milk and I never get a chance to drink it fresh in Britain, so that's what I'd drink.
BARBARA	For me the nicest things to do would be to sleep late – I never get the chance to do that with the children, to have a long breakfast with all the Sunday papers and then to go to the beach for a swim.
ALAN	Yes, my wife. She's my best friend as well. Besides, we've only just got married.
DANIEL	The postman. I really wouldn't want to get any news from home. In fact, I probably wouldn't tell anyone where I was going.
EMMA	The most beautiful place I have ever been is to Delphi in Greece. So I'd go there again and see if it's still as beautiful as I remember.
JANE	Well, believe it or not, the video. You see, I'd want to go with the children, but they're still quite young and very demanding. So if I took a video player and some cartoons on videos, I would at least have half an hour to relax without having to look after them.

Lesson 39　Reading and listening, activity 3

The umbrella man, part 2

My mother was staring down at him along the full length of her nose. I wanted to say to her, 'Oh mummy, he's a very old man, and he's polite, and he's in some sort of trouble, so be nice to him.' But I didn't say anything.

'I've never forgotten it before,' he said.

'You've never forgotten what?' my mother asked.

'My wallet,' he said. 'I must've left it in my other jacket.'

'Are you asking me to give you money?' my mother said.

'No, I'm offering you this umbrella to protect you and to keep, if you would give me a pound for my taxi fare just to get me home.'

'Why don't you walk home?' my mother asked.

'Oh, I don't think I could manage it. I've gone too far already.'

The idea of getting an umbrella to shelter was very attractive.

'It's a lovely silk umbrella,' the little man said. 'Why don't you take it, madam? It cost me over twenty pounds, but that isn't important because I want to get home.'

'I don't think it's quite right that I should take an umbrella from you worth twenty pounds. I think I'd better just give you the taxi-fare.'

'No, no, no!' he cried. 'I would never accept money from you like that! Take the umbrella, dear lady, and keep the rain off your shoulders.'

She took out a pound and gave it to the little man. He took it and gave her the umbrella. He said, 'Thank you, madam, thank you.' Then he was gone.

Lesson 39　Reading and listening, activity 5

The umbrella man, part 4

'He went in that door!' It was a pub. The room we were looking into was full of people and cigarette smoke, and our little man was in the middle of it all, without his hat and coat, and moving towards the bar. When he reached it, he spoke to the barman. The barman gave him a drink. The little man gave him a pound. The barman didn't give him any change. The little man drank it in one go.

'That's a very expensive drink,' I said.

He was smiling now. He went to where his hat and coat were. He put on his hat. He put on his coat. Then, very quickly, he took from the rack one of the many wet umbrellas, and left.

'Did you see that!' my mother shouted.

'Sssh!' I whispered. 'He's coming out.'

He didn't see us. He opened his new umbrella and went down the road. We followed him back to the main street where we met him first, and we watched as he exchanged his new umbrella for another pound. This time it was with a tall, thin man who didn't even have a hat or a coat. When it was over he went off again, this time in the opposite direction.

'He never goes in the same pub twice,' my mother said. 'I expect he's always hoping for rainy days.'

Lesson 40　Vocabulary and listening, activity 2

Just before the Berlin Wall had come down, a man from Leipzig in Germany, who had wanted to be a rock musician for many years, decided to go to England. When he had packed his bags into his old Trabant car, he set off for Liverpool, home of the Beatles, to make his name as a musician.

It was a nightmare journey. He drove all night through Hungary and when he got to the Austrian border he had to queue all day with thousands of other people who had decided to leave their homes and go to the West. He grew nervous while he waited, but the border guards looked carefully at his passport and let him through. After he had entered the West he drove happily across Southern Europe.

He arrived in France and had a minor accident. The crash damaged the exhaust pipe, and the car made more noise than it had done before. At Calais the car broke down and it cost a lot of money to mend it. The ferry ticket was expensive too. He had very little money left now because he had spent so much on the journey. But when he saw the white cliffs of Dover ahead, he felt better.

He left the boat at Dover and he heard a loud noise, so he stopped and got out. The car's exhaust pipe had fallen off. But he headed for London on his way to Liverpool. He was driving along south of London when he came onto the M25 motorway around the capital, which wasn't on his out-of-date map at all. By now, clouds of black smoke were coming from the car, and he was driving noisily in the slow lane. He had driven over a hundred miles around the M25 when the police saw him.

Lesson 40　Sounds

1　He decided to go to Britain.
2　He'd packed his bags.
3　He'd got to the Austrian border.
4　He'd entered the West.
5　He'd arrived in France.
6　He had very little money.

Progress check 36 – 40　Sounds, activity 1

/w/　world　weeks　wash　water　washing　wear
/r/　return　religion　relative　repair　wrapping　remind

Progress check 36 – 40　Sounds, activity 2

/ɔː/　poor　law　more　moor　roar　saw　bore　sure　pour　four　war
/aʊ/　power　tower　sour　hour　flower

Macmillan Heinemann English Language Teaching
Between Towns Road, Oxford OX4 3PP
A division of Macmillan Publishers Limited
Companies and representatives throughout the world

ISBN 0 435 24020 X
    0 333 75055 1 (Turkish Edition)

© Simon Greenall, 1994
Design and illustration © Macmillan Publishers Limited 1998
Heinemann is a registered trademark of Reed Educational & Professional Publishing Limited

First published 1994

Author's acknowledgements
I am very grateful to all the people who have contributed towards the creation of this book. My thanks are due to:
– All the teachers I have had the privilege to meet on seminars in many different countries and the various people who have influenced my work.
– James Richardson for producing the tapes, and the actors for their voices.
– The various schools who piloted the material in Brazil, France, Italy, Spain and the UK: Cultura Inglesa, Belo Horizonte; IBI, Brazilia; CEL Chalon sur Saone; CEL Melun; CEL Chartres; CEL Troyes et Aube; CLM Bell, Bolzano; English Institute, Valencia; San Pablo CEU Monteprincipe, Madrid; Eurocentre, Cambridge; International House, London; King's School, Bournemouth.
– All the readers, especially David Newbold, Peter McCabe and Liz Driscoll.
– Simon Stafford for the stunning design of the book.
– Sue Kay, Janet Bianchini and all the staff and students at the Lake School, Oxford for their enthusiastic trialling of the material, their feedback, support and encouragement.
– Jill, Jack and Alex for providing the most supportive home environment in which to write. Nothing would have been possible without them.
– Jacqueline Watson for researching the photos.
– Karen Jamieson for shaping the project at its early stages.
– Chris Hartley for doing everything a good publisher can do to help.
– Catherine Smith for her kind but very thorough style of editing. Her contribution towards shaping the book, her advice and her attention to detail have played a central role in making this book possible.

Designed by Stafford & Stafford

Cover design by Stafford & Stafford
Cover illustration by Martin Sanders

Illustrations by:
Adrian Barclay (Beehive Illustration), pp26, 50, 70, 85; Rowan Barnes-Murphy, pp62/63; Hardlines, pp41, 53, 87; Martin Sanders, pp12, 14/15, 18, 28, 29, 32, 37, 48, 51, 56/57, 58, 64/65, 75, 77, 89, 101; Jeannette Slater (Beehive Illustration), pp74, 92; Simon Stafford, p25.

Commissioned photography by:
Paul Freestone, pp66, 67; Chris Honeywell, pp38, 39, 42, 43, 60, 61, 78, 82, 83; Simon Stafford, pp68, 73.

Acknowledgements
The authors and publishers would like to thank the following for their kind permission to reproduce material in this book:
British Railways Board and Roald Dahl for extracts from *Roald Dahl's Guide to Railway Safety*; Hamish Hamilton Ltd for *The Kingdom by the Sea* by Paul Theroux; Jonathan Cape for *The Book of Ages* by Desmond Morris; Martin Secker & Warburg for an extract from *The Return of Heroic Failures* by Stephen Pile; Michael Joseph Limited and Penguin Books Limited for *The Umbrella Man* by Roald Dahl; *The Independent on Sunday* for 'Flying is bad for your health' by Jenny Bryan and 'A brief discussion about the environment'; Penelope Mortimer and Hutchinson for 'The Skylight' from *Saturday Lunch with the Brownings*, © Penelope Mortimer, 1960; Virgin Publishing for two extracts from *Urban Myths* by Rick Glanvill and Phil Healey.

Photographs by: Ace Photo Agency/Mauritius p90(b); Adams Picture Library p46; The Bridgeman Art Library p40; The Detroit Institute of Arts, Michigan/Bridgeman p40(m); Britstock IFA pp8(t),68, 69; The J Allan Cash Photolibrary pp32(b), 96; Collections/Paul Watts pp14/15; ©1992 Comstock/Julian Nieman/SGC p21; © Robert Doisneau/Rapho p10; Eye Ubiquitous/J Stephens p86; Fotoccompli p32(t); The Illustrated London News Picture Library pp16, 17, 24; The Hulton Deutsch Collection/Steve Eason p52; The Image Bank p4(t); Images Colour Library pp6, 45, 54(b) Images/Charlie Waite pp22, 23, 34; The International Stock Exchange Library, London p67(tr); Mexicolore/Tito Zauala p80; Courtesy of The National Portrait Gallery, London p40(b); Paramount, courtesy of The Kobal Collection p82; The Photographer's Library p55; Pictor International, London p90; Picturepoint, London pp2, 3, 54; Popperfoto pp3, 94; Reproduced by permission of Royal Mail p78; © Estate of Stanley Spencer 1994 All rights reserved DACS. Stanley Spencer Self-portrait p41; Tony Stone Images pp23, 30, 31, 53; Telegraph Colour Library pp4(b), 8(b); Zefa p44.

The publishers would also like to thank Ella Dennison, Bord Fáilte (Irish Tourist Board), Balvinder Gill, Mrs Jones, Peter Mays, Emma and Kate Simpson-Wells, Coral Stafford, Mr and Mrs Youd and Zoë Youd.

While every effort has been made to trace the owners of copyright material in this book, there have been some cases when the publishers have been unable to contact the owners. We should be grateful to hear from anyone who recognises their copyright material and who is unacknowledged. We shall be pleased to make the necessary amendments in future editions of the book.

Printed and bound in Spain by Mateu Cromo

2003   2002   2001   2000
15   14   13   12   11